SEXUS ANIMALIS

Originally published as *Sexus Animalus. Tous les goûts sont dans la nature*, © Flammarion, Paris, 2020

This book was set in ITC Caslon by the MIT Press. Printed and bound in the United States of America.

Library of Congress Cataloging-in-Publication Data

Names: Pouydebat, Emmanuelle, author.
Title: Sexus animalis : there is nothing unnatural in nature / Emmanuelle
  Pouydebat ; illustrated by Julie Terrazzoni ; translated by Erik Butler.
Other titles: Sexus animalus. English
Description: Cambridge, Massachusetts : The MIT Press, [2022] | Originally
  published as Sexus Animalus. Tous les goûts sont dans la nature
  © Flammarion, Paris, 2020.
Identifiers: LCCN 2021007026 | ISBN 9780262046589 (hardcover)
Subjects: LCSH: Sexual behavior in animals.
Classification: LCC QL761 .P6813 2022 | DDC 591.56/2—dc23
LC record available at https://lccn.loc.gov/2021007026

10   9   8   7   6   5   4   3   2   1

EMMANUELLE POUYDEBAT
ILLUSTRATED BY JULIE TERRAZZONI
TRANSLATED BY ERIK BUTLER

SEXUS ANIMALIS

There Is Nothing Unnatural in Nature

The MIT Press

Cambridge, Massachusetts

London, England

*For Ameline and Marion, my former students,
whom I wish the happiest trails studying
the animal world. I could never have imagined
writing this book without you, funny ladies.*

For Alexandre, whose curiosity is only growing!

Nothing is set in stone.
Everything is diversity.

CONTENTS

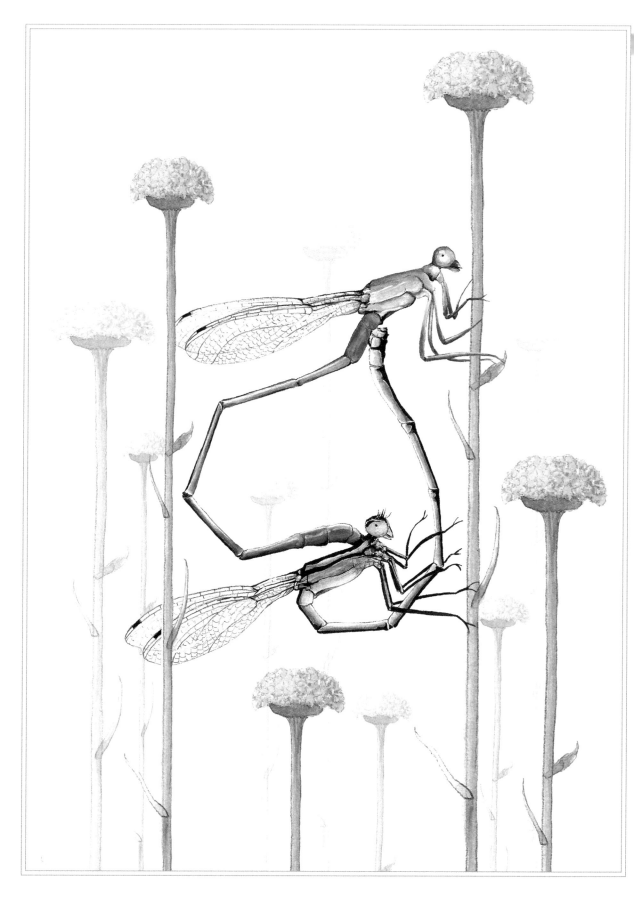

INTRODUCTION

Living organisms have been colonizing the planet for nearly four billion years. They've taken on countless forms through assorted evolutionary and adaptive mechanisms to survive in environments where conditions are highly specific and always changing. As a result, a vast array of adaptive and morphological strategies exist. The whole body—upper and lower limbs, internal organs, bone structure, and tissue—bears witness to this fact. Modes of behavior and biological functions stand at issue, too: patterns of movement, the ways food is ingested, predation, flight, and reproduction. One of the most striking aspects of animals that have evolved to reproduce by means of internal fertilization is the morphological diversity of the genital organs. Remarkably, in species with little variation in terms of general morphology, the differences between the genitals of males (which have received more attention from researchers than those of females) are considerable. It's almost as if natural selection boiled down to just one thing . . . Imagine that!

We now know about gutter penises, double penises, penises that have spines or are shaped like corkscrews, four-headed penises, ones that make sounds, and others that are detachable! Although the data aren't as complete for the female sex, we will duly note examples of vaginas built for storage and clitorises with thorns—exciting findings that promise a wealth of future discoveries.

It seems that multiple reasons underlie this impressive evolutionary development, and the questions that remain unanswered are stimulating. What purpose does the penis serve? Why is it that members of one species have no penis at all, while those of another have two? Why do they vary so much in shape and size? Can they be traced back to multiple points of origin, or is there only one source? Did this happen just to deliver sperm? To increase the chances of impregnation? To ensure exclusive parentage? For survival? And where do vaginas, or the clitoris, fit into all of this? What about pleasure? Does such a thing even exist in the animal world? If so, what forms does it take?

The mechanisms involved are complex. To have any chance of understanding them and providing answers, we need to look at how these organs evolved. That means going back more than 400 million years, to a time well before dinosaurs walked the earth. Until this point, fertilization was essentially external: females expelled their eggs into the water, and males did the same with their semen. But when certain marine creatures emerged from the water, their bodies had to adapt to their new environment. Genitals included: external fertilization won't work if males—and females—launch their sex cells into the air or onto the earth; the biological material doesn't survive for long. Reproduction demanded a real innovation in anatomy—among other things—so that life could thrive on terra firma. That's where internal fertilization comes in: the female keeps her eggs inside her body, and the male adds his semen. To this end, aquatic animals developed pathbreaking solutions: gutter-type penises of the kind still found in sharks, rays, and bony fish (Osteichthyes). Thus, in oceans, rivers, and lakes, many armored fish with jaws—so-called placoderms—equipped themselves with a sexual appendage that is the ancestor of the modern penis. The anatomy of the tiny *microbrachius* proves that internal fertilization dates to the beginning of vertebrate evolution. The first penises were aquatic, then. This mode of reproduction likely proved decisive for the survival of whole species: sperm didn't vanish into the water, and the chances of fertilization improved greatly. In turn, this adaptation proved essential for those vertebrates that abandoned their watery milieu on a daily basis before conquering the continents: the amniotes. By definition, these tetrapods have an amniotic sac protecting the embryo and fetus; their ranks include mammals, birds, crocodilians, chelonians (turtles), and squamates (lizards and snakes). On land, they developed or optimized any number of adaptive features—and, to be sure, changes to the reproductive system. So that the female could keep her reproductive cells inside her body and the male could add his own, most terrestrial vertebrates developed tube-like organs for bringing together eggs and semen. The male organ stiffens through a liquid that can differ between species (blood, lymph . . .) and penetrates the corresponding tracts of the female. The female's genital organs underwent changes, too, but the details remain largely unknown.

Starting to get the picture? It's logical enough, but not that simple. And things are going to get even more complicated. To understand the present, we need to understand the past. Unfortunately, the evolutionary mechanisms that brought forth the penis have hardly been explained in full. Questions persist about the organ's morphology (what it looks like) and its function (what it does). Can we account for this diversity by the fact that the penis evolved more than once? Does the penis have multiple family trees, or did a single, patriarchal penis undergo modifications to produce all others? Let's take a closer look.

Recent research has focused on the origins of the penis and its evolution in amniotes. A "penetrating" organ for copulation was indispensable for prevailing on dry land. The penis ensured the transport of male gametes to the female because air, unlike water, isn't suited for transmitting sperm. One might think, then, that the penis was the only—or, at any rate, the best—solution for reproducing and surviving outside the water. Almost all amniotes that still exist today—mammals, chelonians, crocodilians, and squamates—have a penis or its miniature female equivalent, the clitoris. Almost. In fact, the penis has been "lost" at least two times in the course of evolution: sphenodons and most birds don't have one. The sole living example of the sphenodon—the tuatara of New Zealand—is closely related to lizards and snakes; these animals mate by cloacal apposition (i.e., placing genital orifices into contact), as do most birds. In a process that remains shrouded in mystery, the budding genital organs of sphenodon embryos stop growing before birth! It's as if their genes were keeping alive the memory of an ancestral penis. This, plus the fact that crocodiles and ostriches ("primitive" birds on the evolutionary scale) are equipped with penises, is enough to convince some researchers that the organ has a single origin. According to such reasoning, some lineages developed a penis that others, subsequently, did without. The only problem is that how or why something as important as a reproductive organ could be discarded is hardly clear. It's a good point. Hence other researchers have advanced the claim that penises evolved independently in different bloodlines.

It seems pretty complicated, doesn't it? And we haven't even mentioned arthropods—insects, spiders, and crustaceans—whose copulatory organs also evolved to take on many forms. The names for the penis vary among these animals: mites have chelicerae, spiders have pedipalps, many insects are endowed with an aedeagus, and so on. These modified appendages collect semen to conduct it to the female's genital opening. Anything is possible, but why? The answer is that in a changing environment, sexuality is a matter of survival. When it comes to passing on one's genes, the competition is fierce, both between individuals of the same sex and between males and females. As a result, various species have developed assorted strategies for reproduction, from new organs to outright war between the sexes that culminates in the self-sacrifice of the male. The animal kingdom holds countless surprises in store.

Such strategies are at the root of striking differences in genital form and function. The actual causes, however, remain obscure. Whether origins are multiple or not, fascinating questions persist. How much can shape and size vary? For what reason? To find the right match? To block competition? To get a firm grip? For the purpose of natural selection? Communication? Pleasure? Why is it that some genital organs have a bone? Who do some have spikes? Needless to say, genitals and mating practices in general harbor any number of mysteries. We will dispel a few of them here. Over the course of the book, human organs and sexuality will come to look pretty humdrum. The animal world has us beat on every score.

I

AN UNTOLD VARIETY
OF SHAPES AND SIZES

THE SALTWATER CROCODILE

(Crocodylus porosus)

—

A Gutter Penis

Without a doubt, this is the biggest crocodile species in the world. The largest specimen on record is ten meters long. Saltwater crocodiles grow up in the marshes and freshwater rivers of Asia and Oceania during the rainy season of the tropics; in the dry season, they use currents to reach estuaries and sometimes the sea. This creature is also the most aggressive of its kind. An opportunistic predator both on land and in the water, it feeds on a wide array of other animals (fish, kangaroos, buffalo, monitors, monkeys, dingoes, birds, sharks, and sometimes even tigers or human beings). As a rule, saltwater crocodiles mate in the rainy season. Then, at the start of the dry season, the female digs a hole in the sand, lays her eggs, and covers them up with vegetation. She sticks around to keep watch over eggs and young.

The most distinctive sexual characteristic of crocodiles is the male's penis. Under ordinary circumstances, it remains hidden inside the cloaca—the opening shared by the intestinal, urinary, and genital tracts in birds, reptiles, amphibians, and a few mammals. This organ is perpetually erect, as it were, but concealed. In states of arousal, the crocodile's member doesn't swell or change size. Instead, the muscles holding it in tighten up, and out it shoots; when the muscles relax, it goes back where it was.

Scientists have long had trouble determining the sex of individual crocodiles, and for good reason. To find out, you have to get the organ out of the cloaca to gauge its size. And the trouble doesn't stop there. The female crocodile has a clitoris, but it ranges in length from 90 to 120 centimeters, approaching the dimensions of the penis. Today we know that the two organs are structurally identical in large crocodiles, but the clitoris is smaller; it also stops growing a little sooner.

The crocodile penis usually has a cylindrical shape with slight compressions on the sides, which has earned it the distinction of being called a "gutter." As we noted in the introduction, the first penises in the history of life on our planet looked a bit like this; the genitals of other terrestrial vertebrates are more tube-like. Interestingly, the clitoris has a nonfunctional version of the groove in the penis used for transporting sperm. Documentation as full as this is quite rare, because the female genitalia of most species haven't been studied much. Crocodiles are the exception because already more than 150 years ago, farmers needed to tell girls and boys apart for breeding purposes. It can be a good thing when economic and scientific progress go hand in hand!

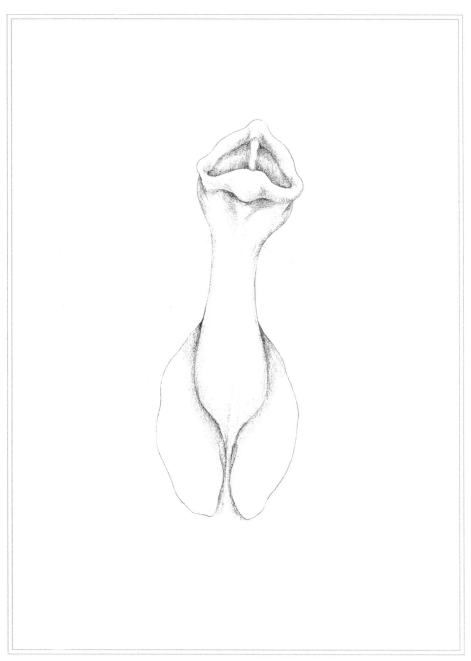

Size of sex organ: about 20 cm (8 in)
Size of animal: about 4.5 m (14.75 ft)

THE OSTRICH
(Struthio sp.)

—

The 3 Percent Club

Let's take a look at what's happening among the birds—and not just any birds. We're talking about a species that originated at the same time as crocodilians, one of the earliest kinds to emerge: Palaeognathae. As everyone knows, this particular lineage has lost the ability to fly. Ostriches! Found only in Africa today, they're the biggest and fastest land birds. Ostriches are monogamous or, when the flock is large, polygamous. Females choose their mate, and then the male attends to incubation practically on his own; he also has the reputation of protecting eggs and chicks. Having experienced several attacks by a certain Coco residing at the zoo in Thoiry, I can vouch for the male ostrich's vigilance. Care of young can last up to a year, and if the danger is great, the female will come and pitch in—along with all her peers! Ostriches are a cooperative bunch.

With such obvious differences, what do crocodiles and ostriches have in common? Both of them have external sexual organs that resemble each other anatomically and serve the purpose of internal fertilization. In simple terms, males have a penis for penetrating the female and delivering spermatozoa. Note that just 3 percent of birds today have this kind of equipment. Ostriches are a rare exception, then—as are other palaeognaths (kiwis and tinamous) and fowl (Galloanserae: chickens, turkeys, megapodes, cracids, and ducks). All other birds reproduce by means of the "cloacal kiss": male and female stick their cloacae together, and sperm passes along to fertilize the eggs.

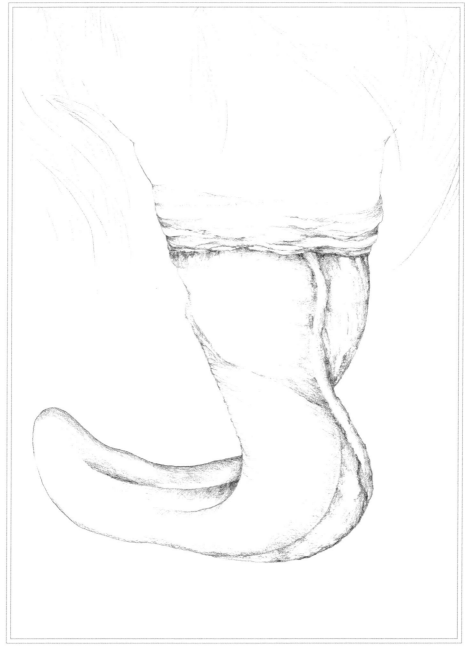

Size of sex organ: about 25 cm (9.75 in)
Size of animal: about 2.5 m (8.25 ft)

Why is it that only 3 percent of birds have a penis? We can reformulate the question. Why is it that 97 percent of them don't have an external penis, whereas most amniotes with internal fertilization do? What could have led to losing an organ that seems vital to internal fertilization—and how can a penis get lost in the course of evolution, anyway? How birds evolved a penis in the first place is extremely complex, and even if genetic history sheds light on these rousing questions, the reason that not having a penis proved advantageous for most birds remains a mystery.

That said, we do have an answer for another long-standing evolutionary question: the lymphatic erection mechanism was likely pioneered by the common ancestor of all birds. For quite some time, no one knew how the largest bird species achieved erection. Now we have determined that male ostriches rely on a stream of lymphatic fluid instead of the system of blood vessels found in reptiles and mammals. But solving one mystery just opens the way for another. The existence of a lymphatic penis in some birds represents an evolutionary puzzle: why was a new structure developed to do something that was already being done? In the case of ducks, lymphatic erection permits the penis to lengthen rapidly and facilitates deep insemination; in ostriches, the lymphatic fluid stiffens the organ by pushing the semen toward the tip. The difference between lymphatic and vascular erections may have led to different mating patterns.

So what's left for us to hold on to? Ostriches and crocodiles have distinct erection mechanisms, but their penises aren't that anatomically different. It may even be that they evolved from the same kinds of tissue. But that doesn't tell us a thing about one of the most perplexing aspects of avian evolution: the vanishing penis. It's only logical to think that evolution would have favored the most efficient mode of reproduction. The loss of an organ as important for propagation as the penis doesn't make intuitive sense. Plus, birds reproduce by internal fertilization! Obviously, we still have much work to do.

THE COMMON EUROPEAN ADDER

(Vipera berus)

—

Not One, but Two Penises

Let's return to the world of reptiles and the crocodile's internal mode of fertilization. The male has a simple penis that comes out of the cloaca through a longitudinal slit; this makes it quite difficult to tell males and females apart. Other reptiles such as snakes and lizards have an entirely different kind of penis—not just in terms of size, but also in terms of quantity. You might doubt what you just read, but that's what it says: crocodiles have a penis that looks like a clitoris, some birds have penises, but even greater surprises await us . . . After all, crocodiles and birds are like people: they only have one penis. How common! Lizards and snakes have two. To be precise, snakes have a bifid penis, which comprises two hemipenises, each with two lobes.

The common European adder, an ovoviviparous and venomous snake also found in Asia, provides a lovely example of these hemipenises. Our illustration shows a hemipenis with its two lobes. Picture that! What is this strange phenomenon? Indeed, this viper has two hemipenises attached to two independent testicles. And if that isn't enough to make your head spin, males alternate between hemipenises for each of the female's vaginas. That's right, females have two vaginas. How practical! What's more, it's not just lymph that makes the male's organs erect (as is the case for birds with a penis), but a combination of lymph and blood. In fact, it's not entirely accurate to speak of an "erection." "Eversion" is the better word, since the hemipenises are inverted when they're in the snake's body. As the herpetologist Marion Segall puts the matter, "It's like turning a sock inside out!" And in evolutionary terms, it's just as fascinating. The penile morphology of snakes is extremely varied. Not only are there different overall shapes and sizes; the organs also display "reliefs" of greater or lesser dimensions. For purposes of comparison, the king cobra (*Ophiophagus hannah*), which is the largest venomous snake in the world, has a member almost 30 centimeters long, and its hemipenises are split into what look like two tentacles. There are as many kinds of penis as there are snakes. Large, small, thin, broad, textured, smooth—even some with bony "claws"! How did such variety arise? What is going on?

One thing for sure is that the highly ornamental morphologies documented by researchers are compatible with a locking mechanism during mating. It would also seem that the various shapes and textures are related to the vigorous bodily twists and turns the female makes. In brief, the more complex hemipenises are in terms of form (including assorted protrusions), the longer the mating will last because of the "locking" that occurs, and the greater the amount of sperm that will be transferred. Is it ever enough? Never. Get a load of this: female snakes, tortoises, and ants are able to store the sperm of different suitors for years! Some of these ladies can even keep their reserves in twenty different pockets to use whenever they want. Human beings didn't invent the sperm bank. Nor were people the first to try to use the best genetic material: the yellow dung fly (*Scathophaga stercoraria*, or, in the vernacular, "shit fly"), which has three pockets for storage, will only use utility-grade semen as a last resort. Yet again, a world of discovery awaits.

Size of sex organ: about 2 cm (0.75 in)
Size of animal: about 60 cm (23.5 in)

THE PYGMY CHAMELEON

(Rhampholeon temporalis)

—

Small Guy, Big Stuff

This tiny reptile, eight centimeters long at the most, inhabits the equatorial forests of Africa, especially Tanzania. One of its close cousins (*Rhampholeon spinosus*) is known for having an especially quick tongue it uses for hunting, which darts out at speeds approaching 90 kilometers per hour in just one-hundredth of a second. Another fascinating aspect of these creatures is the amorous ritual that, when everything goes as planned, culminates in mating. The process of seduction appears quite comical to human eyes. The male must first find a partner who looks like she's in the mood; she should be sporting bright patterns and have a calm demeanor. Then—even though he's much smaller—the intrepid suitor shakes his body and swings his head around his love interest to get her attention. The male's display looks intimidating, even if it's not entirely aggressive. When his performance works, the female will let him climb on her back and then carry him around for a whole day. Mating itself can last from a few minutes to several hours—sometimes all night! It seems that the male does all the work; as is the case with snakes and other lizards, he has hemipenises. The partners join cloacae, and his organs "evert" to penetrate the female.

Once again, we find that lizard penises display great morphological variety. Evidently, the evolutionary diversification in the genital structure of various species is due, among other factors, to the precise way that mating takes place. Chameleons represent a special case because of their wide range of sexual dimorphism. In some species, the males are much larger than females, but in others the opposite is the case. When males are bigger, they will have attributes such as orbital appendages, "helmets," tarsal spurs, and so on. Such visible sexual signs occur when the mating system is determined by territorial males, who mate with several females. These species include iguanid and agamid lizards, as well as many chameleons. Paternity is limited to dominant parties, and the morphological differences between hemipenises are relatively minor.

Conversely, in species where the male is less territorial, smaller, and lacks fancy ornaments—as is the case among anguimorphic lizards (monitors and beaded lizards) and chameleons like our *Rhampholeon temporalis*—anatomical differences are more pronounced. Here females can have multiple partners. Remarkably, genital diversification also depends on mating patterns for many insects (Ephemeroptera, Lepidoptera, Diptera, Coleoptera): when females mate with several males, the rate is twice as high than when they mate with just one partner. The same principle would seem to apply to lizards and perhaps snakes: the less territorial a species is, the more females will "score" with different males—and the greater the variety among penises. It's as if she had to choose penises particularly to her liking. (What does it matter when there's only one option?)

In any event—and to lay this funny hemipenis business to rest—note that anoles have a double penis that develops six times faster than the rest of their body! And when the temperature exceeds 32°C, the central bearded dragon (*Pogona vitticeps*) in Australia changes sex from male to female.

Size of sex organ: about 3 mm (0.10 in)
Size of animal: about 10 cm (4 in)

There's still so much to explore . . . Incidentally, most marsupials also have two reproductive organs; the double penis evidently becomes erect simultaneously, matching the double vagina—and uterus—of females. In kangaroos, for instance, the joey can move from one compartment to another in the course of development, allowing the mother to use the vacant pouch and start another pregnancy. Hard to picture, but there you have it.

THE SHORT-BEAKED ECHIDNA

(*Tachyglossus aculeatus*)

—

A Four-Headed Penis

T he animal world hosts a vast range of penises of varying size and bizarre shapes. Let's leave the birds and reptiles and take a look at mammals. The monotremes are not the least of them. This strange group, according to the most recent classification, includes echidnas (four species) and the platypus—the standard-bearers of Australian wildlife. Their morphology and physiology have fascinated scientists for over two hundred years. Indeed, these creatures are our most ancient mammalian relatives in evolutionary terms. Among other remarkable traits, they stand alone in being both oviparous and lactating; in other words, they lay eggs and nurse their young. The short-beaked echidna (*Tachyglossus aculeatus*) is covered with quills and can roll up into a ball to protect itself from predators. A long, slimy, and very quick tongue serves to capture the insects the echidna loves, particularly ants and termites.

We're supposed to be taking a break from reptiles, but that's not entirely feasible. The echidna—like the platypus—combines mammalian and reptilian features in its reproductive system. For starters, like snakes and lizards, echidnas urinate through a cloaca. They also have an internal penis that's bunched together in a sac in the cloaca until mating time; when it emerges, lo and behold, it turns out to be bifid! But that's where the similarity ends, and a unique feature is evident: each half of the male's penis is divided yet again, giving it a quadruple glans. The urethra extends to all four tips, with many openings to ensure that sperm is distributed as broadly as possible.

A penis with a four-part head! That's a lot, especially since only two parts are ever used simultaneously. How all of this works remained a mystery for quite some time. In concrete terms, the penis has a rotational design, and the two sides alternate in performing their function—which is also the case for snakes and lizards. When erection occurs, the four "buds" of the glans are visible; then, as the process continues, two of them draw back to make room for the others, which, fully engorged, fit the female's genital organ perfectly. Pretty surprising, isn't it? Just wait until you see how mating actually works.

Let's start with the basic facts. Short-beaked echidnas reach sexual maturity between the age of five and twelve. From there on out, they will reproduce once every two to six years. For the species to survive, individuals need to live forty-five years. Here's where it gets racy. The courtship lasts from seven to thirty-seven days. During this period, one to ten males follow the female around, probably attracted by a pheromone. So there they are, ten gentlemen in a row. The "Love Train" is headed for the station. How romantic. The female then chooses a partner. The lucky individual enjoys sexual congress for somewhere between 30 and 180 minutes. (Just for reference, the average duration of intercourse among human beings is less than six minutes.) The amorous pair can be facing each other or have their heads turned in opposite directions, with tails linked; the male often lies on his side, while the female assumes a face-down position. If fertilization occurs, the gestation period lasts between twenty-one and twenty-eight days; the female lays a single egg, which she promptly places in her abdominal pouch. We are getting much closer to the maternal instincts we see in mammals . . . Ten days after she

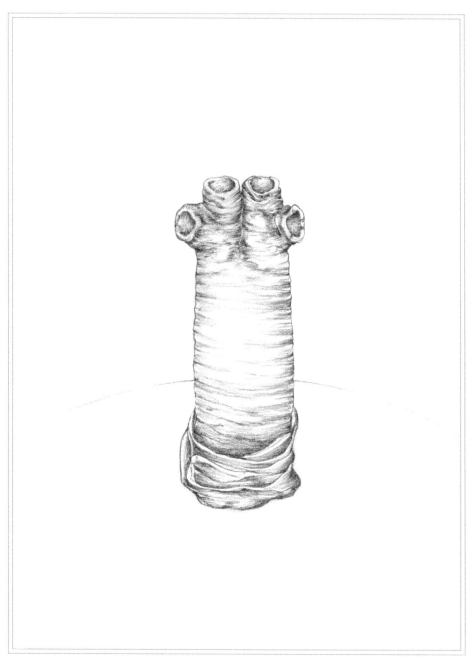

Size of sex organ: about 8 cm (3.25 in)
Size of animal: about 40 cm (15.75 in)

lays the egg, it hatches. The young echidna uses its "egg tooth"—a sharp, white protuberance on the end of its muzzle—to break through; before long, this accessory wears away naturally and disappears.

Newborns are about 1.5 centimeters long and weigh between 0.3 and 0.4 grams. Since monotremes don't have udders, baby echidnas attach themselves to the mother's cutaneous areola, which secretes milk that is pinkish owing to its high iron content. When babies are between two and three months old and have quills, they leave the pouch; at the age of about six months, they stop nursing. Finally, about a year after birth, they leave home and set out on their own. For a story that started with a four-headed penis, that's a pretty wholesome conclusion.

THE TERRESTRIAL
HERMIT CRAB

(Coenobita sp.)

—

The Comforts of Home

Hermit crabs are crustaceans that can live in the water or on land, close to beaches, in coastal areas, and occasionally in forests. The specimen depicted here (*Coenobita perlatus*) dwells in the sand and seaside dunes of the Indo-Pacific region, from Madagascar to Polynesia. These decapods use the empty shells of snails—even stealing them from others—to protect their soft abdomen. As they grow, they change houses, which sometimes means wait lists. Individuals in need of better accommodations form a single line. When the largest hermit crab finds a shell that's right, the others follow suit, big to small, each one in turn. These diminutive creatures have developed symbiotic relationships with other organisms to improve their rate of survival. Thus hermit crabs may be seen carrying sponges on their shells for camouflage and better protection; when they change shells, they take the sponge off the old housing and put it on the new one. Similarly, they can be found living among sea anemones to be safe from predators—specifically fish—thanks to the stinging tentacles. When going about their business, they employ their claws. During my research, I came to appreciate just how effective they were: they're quite skilled at manipulating shells and food and cleaning themselves. That said, I could never have imagined what other researchers have discovered.

Hermit crabs will do just about anything to save their house; it's a matter of survival, after all. Moreover, they take advantage of the protection their shell affords when mating. Once again, it looks like size matters; well-endowed hermit crabs don't even need to leave home! The benefits are twofold: they remain safe during the act of mating, and they don't risk their shell being stolen in the meantime. In effect, mating requires the male to emerge from his shell, at least in part, to fertilize the female. The penis (more precisely, "sexual tubes") discharges a gelatinous substance carrying spermatozoa when placed at the entrance of the female's genital tract. At this juncture, the male is exposed and in a precarious situation. If somebody steals his shell, he can dry out and die within a day. Some of the "penises" of terrestrial and aquatic hermit crabs have gotten bigger in the course of evolution to prevent the theft of house and home during intercourse. In contrast to aquatic hermit crabs, certain terrestrial species modify their dwellings. Their objective is to lighten the shell and make it easier to transport. Then the shell becomes a luxurious accommodation from which the inhabitant is not happily separated. Terrestrial species that refurbish their shells have larger genitals (in relation to body size) than species with more conventional housing. Not all shells are created equal! The most sought-after ones are those that have been customized.

Larger penises are the result of morphological adaptations that enable male individuals to reproduce and, at the same time, to make sure their homes aren't stolen by somebody else. That's safe sex! Private property—the shell, in other words—has contributed to the development of bigger genitals. Unbelievable, but true.

Size of sex organ: about 2 cm (0.75 in)
Size of animal: about 8 cm (3.25 in)

Evolution has generated vast diversity in the sexual organs of animals, and the examples we have covered so far shed light on just a tiny part of a spectrum that includes all manner of shapes, sizes, and behaviors. As we will see, the functions the penis can exercise are just as many—and every bit as unusual.

II

NAVIGATION AND MAXIMUM INTIMACY

THE MALAYAN TAPIR

(Tapirus indicus)

—

A Mobile Penis

The Malayan tapir, or piebald tapir, is the largest of the five tapir species alive today. Easily recognized because of the white "saddle" on its back and the tips of its ears, it is the only Asian tapir known to exist. These solitary creatures mate in spring. As a rule, females give birth to just one young every two years, after a gestation period of just under four hundred days. The Malayan tapir sports a prehensile trunk, but its trunk is not its most improbable feature. Indeed, the tapir does a good job of hiding its real trump card, because the organ in question manages to move in any number of ways . . . Needless to say, I'm talking about its penis.

Like dolphins and, as we will see, elephants, Malayan tapirs are endowed with an enormous penis that's extremely agile and ideally suited for getting the job done. You might not think so just by looking at him: his penis is folded up, like an accordion, in a sheath inside his body. The glans has a

Size of sex organ: about 90 cm (3 ft)
Size of animal: about 1.80 m (6 ft)

distinctive shape, like a mushroom. But the truly remarkable feature—and efficacy—of the organ is how it manages to find just where to go. A penis that can move on its own and find the genitals of the female is a real asset when it comes to wooing the ladies and ensuring successful reproduction.

The gigantic and enigmatic organ of generation the tapir possesses has given rise to some pretty wild folklore; some admirers speak of the "phallic spirit of the forest." But the story has a sad side. The tapir's reproductive habits are receiving the attention of researchers today because habitat loss is paving the way for the creature's extinction. It seems that fewer than two thousand Malayan tapirs live in the wild, and about two hundred in zoos across the world. Unfortunately it is extremely difficult to get them to breed in captivity. I'll never forget when little Tengah first arrived at the Jardin des Plantes, in Paris. He dazzled everyone, gamboling about in his new home (sometimes a little clumsily) in front of happy children. Ever since a girlfriend has joined him, there's more hope that the species will survive.

THE AFRICAN BUSH ELEPHANT

(Loxodonta africana)

—

A Second Trunk

An adult male African savanna elephant stands almost four meters high at the shoulder and weighs up to six tons. His female counterpart reaches a height of three meters and weighs about four tons. Elephants achieve sexual maturity around the age of fifteen, but males don't start breeding until they're over thirty—and strong enough to fight other bulls for cows who aren't pregnant or raising young. The twenty-two-month gestation period is the longest of all the mammals, and baby elephants can nurse for up to four years. On average, elephants have one sexual encounter every five years. That's enough to make males a little pushy. During musth—that is, when the animal is in rut—testosterone levels increase to fifty times the normal level. Males swing their ears, shake their heads, and whip out a penis two meters long, as imposing as it is surprising. As if elephants weren't already fascinating enough, they have a second trunk! Even the tapir pales in comparison: not only is the elephant's penis massive, it's prehensile. You heard right . . . One of David Attenborough's celebrated documentaries shows a large bull using his stuff to stand up, swat flies, and scratch his belly. Indeed, it would seem that elephants can use their penis to grab leaves, fruit, bark, or pieces of wood. Incidentally, an elephant clitoris is about 40 centimeters long.

But do elephants really need a second trunk to mate? You'd better believe it. For an animal that weighs six tons, it's useful to have a prehensile penis. Mating would be practically impossible for a creature so gigantic without some means of assuming the proper position for performing the rhythmic thrusts necessary to do the deed. Since bull elephants are extremely heavy, copulation occurs very rapidly, generally lasting between twenty to thirty seconds. If I may: that's not very long, so the male should really get straight to the point. A prehensile penis is just the thing. Nor is that all. During the mating season, bull elephants produce extremely smelly, dark-green urine that they often mix with mud, then roll in or spray on the ladies to win them over—what charmers. My "elephant research team" in Namibia and South Africa studying how elephants use their trunks observed one of them masturbating with it.

But when it comes to oversize equipment, forget elephants—and even whales. The animal with the most imposing organ is minuscule in comparison: a tiny, hermaphroditic crustacean just a few centimeters long. Barnacles have reproductive organs eight times their own length. That's pretty handy, since these little creatures are attached to rocks and can't move. To explore their surroundings and find a partner, barnacles use their penis. And if they can't find a mate, they have another option: self-fertilization.

I can't believe it never occurred to me before: the largest sperm in the world belongs to a fruit fly! *Drosophila bifurca* produce spermatozoa that are all folded up—but once unrolled, they're twenty times the animal's size: six centimeters. If the same proportions applied to human beings, that would mean sperm over thirty meters long.

A giant animal, the elephant, is credited with having the biggest penis. And the fruit fly has the biggest sperm. Go figure . . . that's evolution. Plus, large animals ejaculate at high speed, and small ones do so at low speed. What does it all mean? Large spermatozoa compensate for low-powered ejaculations, while small spermatozoa enjoy a soft landing even though they're discharged rapidly: big ones are given the chance to find the egg even though they're launched slowly, and little ones manage to avoid being crushed, even though they're blasted at a breakneck speed.

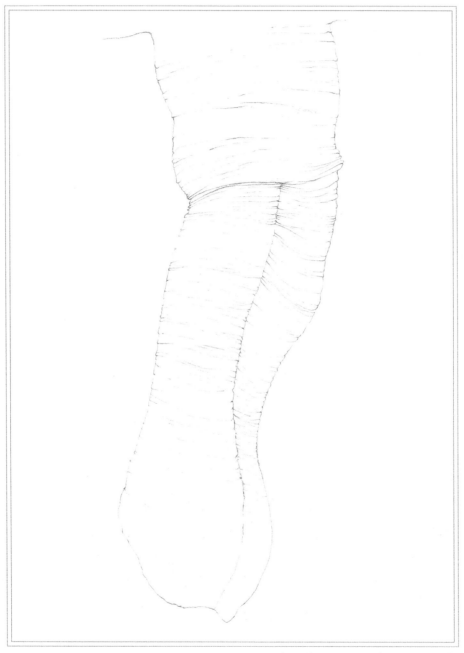

Size of sex organ: about 2 m (6.5 ft)
Size of animal: about 4 m (13 ft)

THE GREATER ARGONAUT

(*Argonauta argo*)

—

A Detachable Penis

Whereas barnacles have developed a long penis because their bodies are immobile, argonauts have adopted another strategy: they detach their sexual organ. A few preliminary remarks are in order.

The argonaut is a cephalopod mollusk, a close cousin of the octopus and squid. By means of her front tentacles, the female secretes a thin, white shell of calcium carbonate in spiral form. Only females do so, since the shell is meant for storing and protecting eggs. Incidentally, the thin membranes of this structure have been compared to sails—which is why Carl Linnaeus named the mollusk "argonaut," in reference to the Greek sailors of legend. The female's conspicuous feature heightens the marked sexual dimorphism of these creatures—to say the least. Adult males are just 1 or 2 centimeters long, while females reach 40 to 50 centimeters. So the women are ten to fifty times bigger than the men. My heart goes out to the little guys. How is reproduction possible when there's such a disparity in size? Well, the male has found a remarkable solution.

These mollusks are equipped with eight tentacles. But the third one on the left is huge because it performs a different function from the others: in fact, it's been customized for copulation—it's an arm that plays the part of a penis! This so-called *hectocotylus* consists of a reservoir for spermatophores and a thin tip, like a penis. An arm that stores and transmits sperm—and that's not all. Let's provide a few details. Male argonauts transmit spermatophores to the female by inserting their hectocotylus in her copulatory orifice. This agile appendage can grope around for hours in the female's mantle in search of her pallial cavity. That's all the contact between partners that occurs during mating. But what unique contact it is: the male's penis, at the end of the hectocotylus, separates from his body once it finds its way into the female's cavity. A detachable penis: there you have it. It's difficult for human beings to believe—so difficult, in fact, that the great anatomist Georges Cuvier thought he'd found a parasitic worm when he discovered one in a female specimen.

It would have been better to trust Aristotle's authoritative pronouncement in *History of Animals*. This was no worm, but a mating arm. The hectocotylus grows underneath the left eye of the male argonaut; it emerges when he reaches sexual maturity. Just as remarkably, females store the spermatophores until they are ready to lay their eggs in clusters on the "branches" at the bottom of their "basket." As a rule, males die in the months after mating; sometimes the female will devour her paramour. The females guard their eggs, which bring forth embryonic argonauts that develop for a few weeks inside the structure before the young finally emerge, at night. Shortly afterward, the mother dies. Here's another remarkable fact: multiple organs from different suitors have been found in a single female. Do the ladies choose the best spermatophores?

THE MALLARD
(Anas platyrhynchos)

—

The Corkscrew Rapist

Ducks' penises are made to get as close as possible to the eggs so as to optimize fertilization—to really get in there. Their members become erect thanks to lymph, not blood. Scientists have yet to determine the origins of this mechanism, but it might have something to do with a kind of evolutionary "arms race." To be blunt, male ducks often rape females. The goal is to spread their genes. The means to this end is lymphatic erection, which allows the penis to lengthen rapidly and facilitates deep insemination. Females try to prevent this from happening. An explanation is in order.

Almost all birds mate by means of the "cloacal kiss": the male climbs on top of the female and presses his cloaca against hers to transmit his genetic material. Evolution has unfolded so that almost 97 percent of birds do not have a penis; why this is so remains a mystery. However, a few species among the geese, swans, ducks, ostriches, rheas, and other emus have external penises that develop in the spring, only to disappear until the following year. For a good number of females, who will never know the tenderness of cloacal coupling, the mating season turns into a nightmare.

So ducks are part of this minority. Some do have a peaceful sexual life, free from violence. But not all of them. Everyone, at one point or another, has gone to the park to admire the mallards and maybe thrown them a bit to eat; they look cute and affectionate, walking around in pairs. But behind this decorous exterior, the males are real brutes. During the breeding season, competition is fierce, especially given that the males outnumber the females. Tensions run high, and shocking behavior ensues. These ducks have developed several means for maximizing their chances of reproduction, each one worse than the next in our eyes.

The first is gang rape. Evidently, half of all female mallards experience sexual assault. And as it turns out, rape isn't the province of ducks alone. Among others, there's the sea otter (*Enhydra lutris*)—that adorable little maritime killer. When male sea otters can't mate with females because of their dominant peers, they get frustrated and "take what they can get." These frustrated males even rape baby seals, holding the victim's head underwater for more than an hour and a half—which can obviously have fatal consequences. And while we're on the subject, sexual assaults by seals on penguins have been documented in Antarctica. Is this some kind of revenge? Probably not. It would seem that the shrinking territory available to large populations makes the young males witness mating activity that they're eager to try out for themselves. The preferred euphemism among researchers is "sexual coercion"; whatever one calls it, it's not consensual. Finally, penguins can be sex offenders, too. Male Adélie penguins (*Pygoscelis adeliae*) have been observed copulating with injured females, chicks who have fallen out of their nest, corpses, and other males. Add up all this shocking behavior (and there's probably a lot more we don't yet know of), and our anatine rapists don't seem that unusual.

Besides rape (which, unpleasant as it is from the human standpoint, serves to optimize reproduction among mallards), the males have another resource for achieving impregnation: their corkscrew-shaped penis, which facilitates a forced entry into the female's sexual organ. Here things get really interesting, because the females have developed ways to defend themselves. The female mallard has a twisting vagina—a kind of maze going in the opposite direction of the male's penis to prevent penetration by force. And the story doesn't stop there. The war between the sexes has escalated. The male's

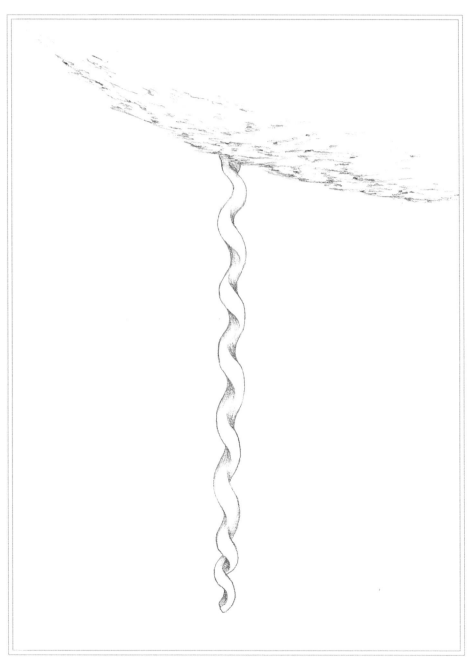

Size of sex organ: about 20 cm (8 in)
Size of animal: about 55 cm (21.5 in)

evolutionary response is to have developed a penis matching the length of the vagina. Moreover, the more competitors there are for a female, the more the males will pump up their equipment. Among ruddy ducks (*Oxyura jamaicensis*), an aggressive lot, dominant males in the so-called explosive period of erection sport penises up to 20 centimeters in length, which is sometimes even longer than their bodies. Intoxicated by testosterone that causes testicles to swell from 1 to 125 grams, the male deploys his erection in less than half a second and takes the female by force. Some penis sizes really beggar belief. The South American lake duck (*Oxyura vittata*) sports a penis 42.5 centimeters long, once the corkscrew is fully extended. That's something else for a creature who's barely 40 centimeters from one end to the other. It might represent an advantage for reproduction, but still quite the ordeal for females of the species.

It's a trial, but the female ducks don't give up easily. Since coiled vaginas aren't enough, some have managed to block the sperm of undesirable partners. How, you ask? By means of little valves! Female water striders (Gerridae), those little insects on ponds, can also seal off their vaginas. Some spiders in the Theridiidae family—which weave those tangled webs—have a labyrinthine vagina that loops around so sperm won't reach the eggs; if the female decides she wants it, she opens an internal sluice to let the sperm through. Some mollusks have mastered these tricks, as have dolphins, whose vaginas include folds that can limit penetration; the female makes subtle movements during copulation to send the penis to the wrong side of the vagina and prevent fertilization. That's a real "rhythm method." And the surprises keep coming. Certain species of butterfly, for example, have "vaginal stomachs" (*bursa copulatrix*). These specialized organs digest sperm that has been stored—and rejected. The semen is rich in protein, so she gets something out of it, after all.

THE LEOPARD SLUG

(Limax maximus)

—

A Martyred Penis

Reaching a length of up to 20 centimeters, the leopard slug is a gastropod that lives in the temperate zones of central Europe, in wetlands, near rivers or streams, and in forests and gardens. It eats plants, mosses, mushrooms, dead wood, and occasionally members of its own species who have died. How delectable. The creature owes its name to the brown spots scattered all over its body. But these aren't the reasons for its renown. What earns the leopard slug its fame are its performative antics, showing off a blue reproductive organ as big as its body!

First off, a reminder: most gastropods are hermaphrodites—that is, they possess both male and female genitalia. This feature is quite practical. Snails, for instance, can produce eggs and sperm all on their own, even though they seek out partners. The act itself isn't very glamorous in most species, however. For many gastropods, "Cupid's dart" is essentially a calcareous stinger violently planted in the partner's flesh. Classy.

Among slugs and snails, the same individual is male and female at once. More precisely, the leopard slug (*Limax maximus*) starts out male and, under the influence of specific hormones, becomes female. Some of them can even fertilize themselves and reproduce without knowing the joy of sex—or, at any rate, what passes for it. For this slug, mating represents a veritable feat. The acrobatic gastropods climb a tree or some other vertical feature of the landscape, then embrace each other while suspended in the void on a sticky strand of mucous. Breathtaking! What a way to spice things up. Squeamish parties might want to avert their gaze at this juncture. Now, an enormous protuberance extends from behind the head: the penis. Both partners drop their penis down and wrap around each other. Gravity makes the enormous appendages swell and optimizes fertilization. The slugs exchange sperm at these extremities; then the penis is drawn back inside, and actual fertilization occurs. But the sight is truly sublime: the genitals are bluish white, the color of the fluid circulating in their bodies. After mating, the slugs lay about two hundred eggs each, which hatch between twenty and more than forty days later, depending on weather conditions.

The poetry—which is relative to begin with—stops there. In some species of slug, the penis can measure twice the animal's length. (Incidentally, the banana slug [*Ariolimax dolichophallus*] has a penis that can reach 30 centimeters; it's one of the biggest organs relative to body size.) Worse still, the organ can sometimes get stuck in one's partner (after mating that can last for hours). In such cases, the inconvenienced slug has no qualms about gnawing on the penis and ripping it off. This delightful phenomenon is called *apophallation*. Deprived of manhood, that slug will henceforth play only female roles.

Size of sex organ: about 30 cm (12 in)
Size of animal: about 15 cm (6 in)

III

COCKBLOCKING

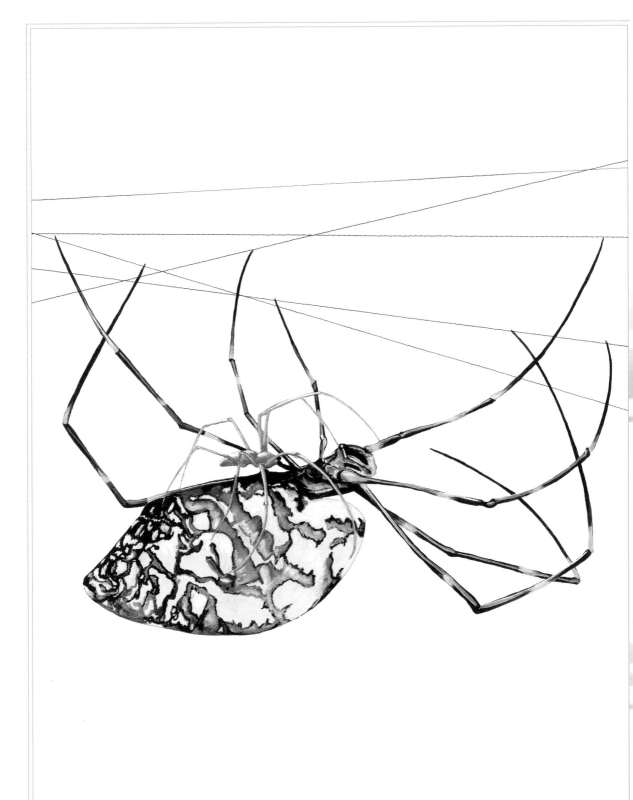

THE GOLDEN
ORB WEAVER

(Nephila sp.)

—

The Ultimate Sacrifice

Having spent our fair share of time on penises of unsuspected size and shape and mobile units designed to find their way under difficult circumstances, let's take a gander at the various means of preventing rivals from doing the same. If the goal is to ensure reproductive success by fertilizing as many eggs as possible, many males try to mate with as many females as possible.

Conversely, a good number of females are polyandrous, mating with several males. Some females, to be found among insects, mollusks, and spiders, avail themselves of a *spermatheca*: an organ that keeps sperm available for future use. "Sperm competition" (a technical term, by the way) is quite intense, and it's closely tied to adaptive strategies for ensuring paternity. The goal is to increase the chances of achieving fertilization without some other guy sidelining you. But how is it possible to get rid of a rival's sperm when the female has already stockpiled it? How do you keep her from mating with a competitor? What can you do to maximize the chances of success for your own spermatozoa? Many insects—including mosquitoes, butterflies, bedbugs, and other beetles—transmit antiaphrodisiac substances that inhibit the sexual receptivity of the female after they're done. The simplest thing is to make the ladies repellent. In somewhat more gracious terms, the males force the females to be monogamous by taking away their power of attraction.

But some species go even further. What better example, for such a charming subject, than . . . the spider. It's actually quite unfair that spiders are so disliked. I, for one, appreciate them more and more each time I attend a talk by my colleague Christine Rollard. Spiders are extraordinary beings, and most of them are just magnificent. *Nephila clavata* confirms the rule. No one will deny that the behavior this creature exhibits is impressive, if nothing else. Among arachnids, reproduction can be cruel. Cruel? Yes, indeed, even if it starts harmlessly enough. First, the male masturbates, then he gathers his semen in his pedipalps,[1] where his sperm is stored in the final segment, the tarsus (or palpal bulb). At mating time, he uses his pedipalps to deposit his sex cells in the female's spermathecae, which are located under the abdomen.

The principal organs involved in this kind of mating are the pedipalps, then. As Rollard observes, these "paws" look like boxing gloves, and they're vital for telling males and females apart. In some species—including *Nephila*—the males put their pedipalps inside the vaginal opening and then leave them there by cutting them off! That's one way to beat the competition: leave your copulatory organ in your partner's epigyne (genital aperture) so that there's no room for whoever else comes along. Admittedly, it's a little gruesome. And here's a fun fact: depending on the species, each male has a different bulb that gets inserted into the female's organ in varying ways. Whatever the details of individual technique may be, insemination occurs, and the chances of fertilization improve. Hey, it works . . .

So personal style can differ, but the objective remains the same: to keep other males from fertilizing the coveted female. Just to be sure, the party

---

1   Appendages on the male's head, performing much the same function as insects' mandibles.

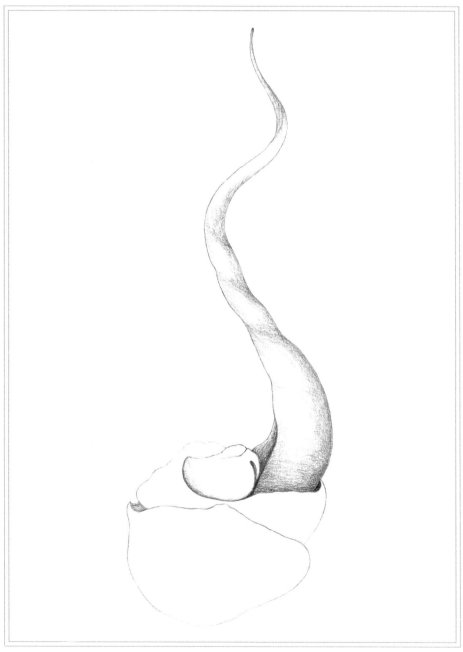

Size of sex organ: about 0.8 mm (0.3 in)
Size of animal: about 5 cm (2 in)

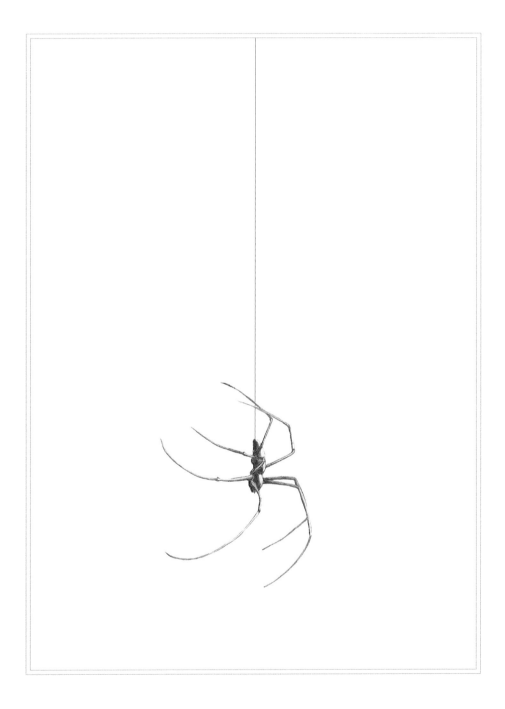

who has emasculated himself may also stay on the web and attack unwanted interlopers. *Nephila* males who have parted ways with their bulb are more aggressive than those who are still fully endowed. It's understandable. But there's an issue at stake other than siring offspring: basic survival. It so happens that some female spiders are known to practice cannibalism. Sexual dimorphism is sometimes quite considerable, and the male is tiny compared to the female. He leaves behind his palpal bulb, and insemination proceeds on its own; meanwhile, he can take off and gain safety (or at least try). Jettisoning the reproductive organ may have developed in response to murderous behavior on the part of females. If such is the case, let's wish the little guys success.

THE DRONE

(Anthophila sp.)

—

Corking the Queen

Male bees are drones. In contrast to female bees, they have no stinger and do not collect nectar or pollen. The drone hatches from an unfertilized egg and mates with a fertile queen in a "nuptial flight." That's all well and good. But how does the process unfold?

The drone's penis is internal—a so-called *endophallus*. The stakes of reproduction are high for males, because with few exceptions, they do not survive mating; moreover, the queen will mate with five to twenty of them. As a result, the drone needs to make the most of his only chance to embrace Her Majesty! To do so, he approaches from above and straddles her; he grips her with all six legs and the small hooks at the base of his genital organ. He needs to hold fast because mating occurs while flying, ten to forty meters up in the air.

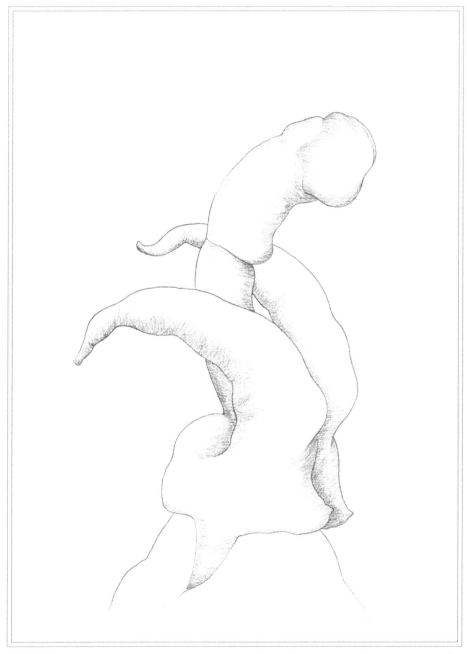

Size of sex organ: about 3 mm (0.125 in)
Size of animal: about 1.2 cm (0.5 in)

Once the couple is united, the male's abdominal muscles increase the pressure of the hemolymph and "inflate" his penis, which everts and turns inside out (like a snake's). In this state, it can penetrate the queen's body for one to five seconds maximum. The process is fast but effective, since a large quantity of seminal fluid and sperm is ejected at a very high speed and with great force. Ejaculation is so explosive that it hurls the drone backward, away from the queen. In fact, the endophallus breaks in two during the violent separation, and half remains with her. That is really the height of indignity. In a fraction of a second, the sperm enters the fold where the queen has her stinger before reaching the oviduct, making a distinct "pop" sometimes audible to the human ear! After a feat like that, the male doesn't survive long; meanwhile, his penile bulb serves as a plug to thwart those who come along after him. They will try their luck, anyway; we know that the queen loads up on almost 100 million spermatozoa, from several males. But the semen of the first suitor has the advantage, since it's trapped in the oviduct, so to speak. Unused spermatozoa are stored in the spermatheca for fertilizing Her Majesty later in life. Once this treasure has been used up, it's time for another queen bee.

THE SEA SLUG

(Chromodoris reticulata)

—

Penis Wars

Sea slugs are nudibranchs. These marine gastropod mollusks are distinguished by their lack of shell, hence the designation "slug." Each species of nudibranch is more magnificent than the next, a real marvel. But looks aren't why we're talking about these sublime creatures here. Like other gastropods, sea slugs are hermaphroditic and endowed with both a male and a female genital system. In fact, the male and female organs function at the same time. When no partner is to be found, these slugs can generate an egg on their own and reproduce by parthenogenesis. Animals as varied as sharks, reptiles, amphibians, fish, and numerous species of insects (aphids, for example) can produce young independently. When sea slugs do mate, the surprises—which aren't necessarily pleasant—are proportionate to their beauty.

First, the prospective partners draw close to each other. In the case of *Chromodoris reticulata*, a species about six centimeters long, the individuals assume a position side by side on the rocky floor of their native Pacific Ocean. Without further ado, they launch a penis war to intimidate each other—specifically, they place their male organs in contact with each other and perform symmetrical movements. This is quite an original way to indicate amorous intent. And that's just the beginning. For a few minutes at most, the penises penetrate the corresponding female orifices of the other slug. The goal is to transmit male sex cells to the other party so as to fertilize the female gametes located in an internal pocket of their bodies. Small spines pointing backward adorn the end of the penis. And why is this? The spines serve to expel spermatozoa collected during a previous encounter. In this manner, the slug ensures that only its own genetic material will fertilize the partner, not someone else's. Yet again, we bear witness to sperm competition that takes the form of mutilation. Whether there's any real competition or not, the sea slug leaves its penis behind, and a new one grows back in just twenty-four hours. The prodigious organ is extremely long and coiled up in a spiral inside the slug's body—so it's actually just the tip that breaks off. Sounds like magic, but it's scientific fact.

To conclude on a softer and more poetic note, when both parties are fertilized, they each carry the eggs. When the eggs are laid, we behold another wondrous sight: a coiled ribbon made up of tiny colored beads.

Size of sex organ: about 5 mm (0.2 in)
Size of animal: about 6 cm (2.25 in)

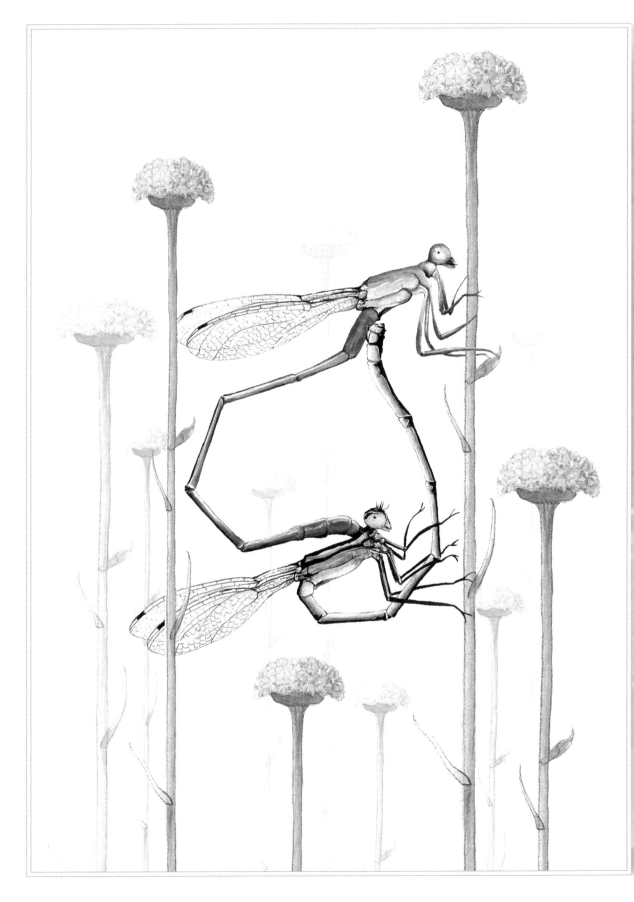

DAMSELFLIES
(*Zygoptera sp.*)

—

Evicting Sperm

"**D**amselfly": what a pretty name. It might sound charming, but it conceals unexpected cruelty. Among many insects, the morphology of the penis is perfectly matched to the internal anatomy of the female's genitals. Unfortunately, this doesn't prevent males from expressing themselves to the detriment of their partners. Male damselflies—like dragonflies, a close relative—exhibit astonishing anatomical peculiarities.

To appreciate just what is happening, it's important to understand the act of mating itself. Most of the time, the male straddles the female, though sometimes the female rides on top of the male. But they can assume any number of other positions in keeping with anatomical constraints. What's remarkable about damselflies is that they produce sperm at one end of the abdomen, while their copulating organ is located at its base. Hardly a practical design. To remedy the situation, the male, before mating, bends his body to fill up the reservoir that is part of his reproductive apparatus with sperm. During the act itself, he grabs his partner between his head and thorax by means of the pincer at the end of his abdomen. His difficulties don't stop here, though. The female's genital opening is at the end of her abdomen, so she needs to bend forward for contact with the penis to work. Damselflies form a heart-shaped figure when mating. That might sound poetic, but things get pretty ugly.

The competition is tough. When damselflies set about reproducing, a whole gang of males gathers at places suitable for laying eggs. Very few of them will manage to mate with a female. What's more—as is the case with other arthropods and mollusks—female damselflies are endowed with a spermatheca and can hook up with several partners, then save the sperm for fertilizing eggs later on. While this may be a boon for the female, it poses a danger for males, whose sperm can get crowded out by somebody else's. Consequently, male damselflies have a strategy of their own for optimizing the transmission of their genes. Evolution has equipped them with a reproductive organ—the aedeagus, to use the technical term—that's shaped like a spoon. That's right. And the male uses this mighty appendage, a veritable work of art, as a scraper to remove the sperm of his predecessors. Some dragonflies even have thorns, bristles, or other barbs on their organ. So for a few minutes when the female is, shall we say, busy, the male is hard at work scraping out between 90 and 100 percent of the sperm left by his predecessor. Then, in just a few seconds, he makes his own deposit in the sperm bank. Whoever does so last basically gets to be the proud father. Kicking out rival sperm to ensure the greatest chance of fertilization—somebody was bound to do it.

To conclude our discussion of the matter, note that suppressing rival sperm can also occur by chemical means. Some mosquitoes have a substance in their semen that will act to expel any sperm that is added later. By the same token, but in reverse, some members of *Drosophila* have sperm that will drive out anything rivals have already left behind.

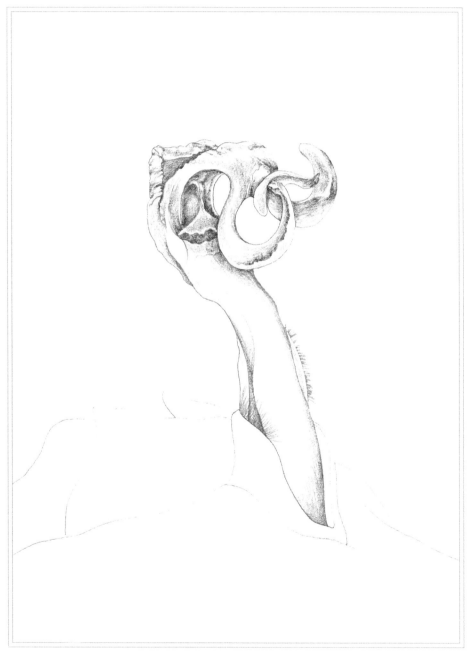

THE PHASMID

(*Necroscia sparaxes*)

—

All In

Shutting down the competition is a veritable preoccupation for males eager to ensure the transmission of their own genes. As we have seen, some animals detach their penis, and others use a mating plug (or *sphragis*); meanwhile damselflies take the liberty of removing rivals' sperm. Needless to say, other strategies exist as well. If I may: that's what is so much fun in the animal world—anything is possible, and further surprises always await us.

So there you are, a solitary male. There's no guaranteeing that you'll pass your genes on to your dream girl. You're apprehensive about not having any progeny. Let's say you can't remove your penis, and you also lack a tool for scraping away the sperm of those who came before you. That's a handicap, to be sure. Things are off to a bad start. Well, you could follow the lead of the *Necroscia* genus, and *Necroscia sparaxes*, in particular. The male of this species of stick insect sticks around in the female's genital tract, even after ejaculation. Sound like a good idea? Be careful: we're not talking about five minutes. The male *Necroscia* remains in intimate contact with "his" female for hours or even days. In fact, in one case on record, the male's tenancy lasted a mind-boggling seventy-nine days! There's probably a golden mean, but one can't deny that the strategy is effective. No one else can show up and interrupt fertilization.

Along similar lines and with the same goal, males of certain species of bedbugs like *Parastrachia japonensis* remain attached to females for an hour after mating. The Colorado potato beetle (*Leptinotarsa decemlineata*) mates several times in a row with the same partner: a practical way to ensure you are the only one filling up the spermatheca and eggs fertilized in the future will come from you alone. For other species, it's necessary to keep the female busy to mate for an extended period and as effectively as possible. For example, among nursery web spiders (*Pisaura mirabilis*), which are common in Europe, the male offers the female an insect in silk "paper." Not only is this a way of winning her heart; it also serves to distract her so that mating will last longer: she has to unwrap the gift. While she does so, the male holds her in his embrace—and then makes himself scarce to avoid becoming a meal. By the way, spider suitors are known to pull tricks on the ladies: sometimes the present is all wrapping, and there's nothing inside!

To conclude our review of strategies for foiling the competition, let's look at a few more examples. Mating plugs, we noted regarding the drone, do the trick quite well. Species other than bees use plugs too. Some cockroaches (*Blattella germanica*) and migratory locusts (*Locusta migratoria*) inject a biological cement that hardens to block the intromission of another penis. Butterflies shoot a viscous secretion that assumes the shape of the female's genital tract and then solidifies. Finally, other animals sacrifice their genitals—including the male *Dinoponera quadriceps* ant, whose genitals remain stuck in the female. There are many ways to beat the competition, then. When it comes to ensuring paternity and optimizing fertilization—as in all spheres of life—animals are endlessly resourceful.

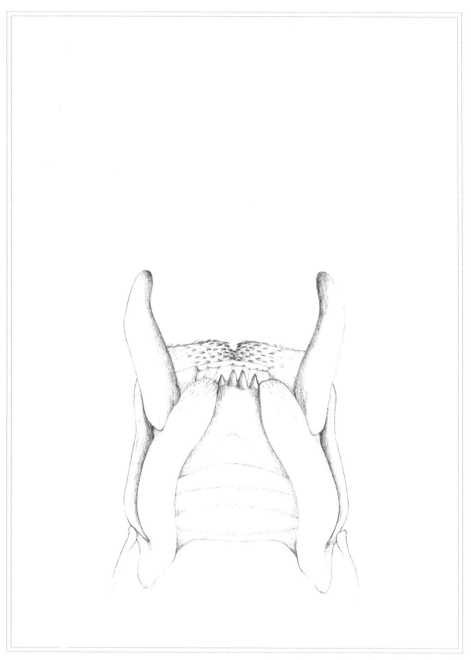

Size of sex organ: about 5 mm (0.2 in)
Size of animal: about 8 cm (3.15 in)

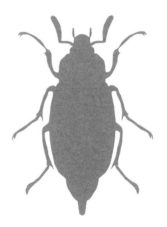

IV

HOLD FAST
AND DRILL DEEP

THE WALRUS

(Odobenus rosmarus)

—

The Biggest Boners

As we've noted, there are various ways to optimize fertilization. One of them, which is common in mammals, involves . . . having a penis bone. That's right. You might not have known it, but some of these organs are equipped with one. The highfalutin scientific term is *baculum*. In the vernacular, we call it a "love dart."

Many animals—most primates, as well as bears, moles, shrews, bats, hedgehogs, cats, dogs, seals, otters, and raccoons (among others)—have a penile bone. Its function is to prolong the intromission of the male organ in the female genitals, in order to prevail over rivals and increase the chances of siring young. Accordingly, species that do the deed the longest also have the longest baculum. For reference, the smallest penile bone belongs to a tiny monkey, the marmoset, equipped with a two-millimeter dart. At the other end of the spectrum, the walrus holds the record, with a bone that's 63 centimeters long. Yes, indeed, the walrus sports the heaviest artillery among mammals today relative to size. A fossilized walrus baculum dating back twelve thousand years has been discovered in Siberia that—hold on to your seats—measures 1.4 meters in length.

Another point of interest is that polygamous primates sport a bigger boner than monogamous ones. Why might this be? Because polygamy implies an extremely high level of competition between males—especially when the mating season limits the time available to find love. A big baculum presents a distinct advantage. Clearly this principle does not apply to human beings, who (apart from rare cases documented in medical literature) do not have a penile bone. It appears that 20 percent of human societies are monogamous, while the rest are polygynous, and competition between males is hardly lacking. So is this bone quite as important for reproductive success as has been claimed? The matter awaits further discussion.

For one thing, we too readily forget that mating can also be a source of pleasure. Second, the fact is almost systematically neglected that the females of many species possess an anatomical equivalent: the clitoral bone, or the *baubellum*. (Incidentally, the very presence of a clitoris in animals tends to receive no attention.) Then again, in many species, neither females nor males have a bone down there; besides human beings, examples include monotremes (echidnas and platypuses), marsupials (kangaroos, possums), and spider monkeys. Why is this? According to some scientists, the disappearance of the genital bone could be an effect of neoteny (the retention of juvenile characteristics in adults), characterized by an incomplete skeleton. It's possible that chimpanzees have penile and clitoral bones, and we do not, because we are born in a more immature state. And, I would hasten to add, that's a good thing! If babies were entirely "finished" at birth, their heads would be too big to make it through the narrow, bipedal pelvis of adult human females.

The presence or absence of these bones remains a mystery—as do their origins. Among mammals, it would seem that the appearance of the love dart marks the point of evolutionary separation between the placental and nonplacental lines (that is, "true" mammals, on the one hand, and marsupials and monotremes, on the other). To be precise, it is supposed to have developed before the most recent ancestor shared by carnivores and primates,

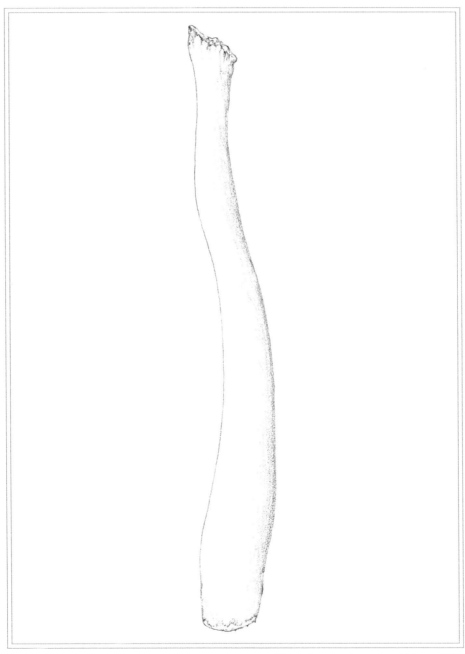

Size of sex organ: about 90 cm (3 ft)
Size of animal: about 3.5 m (11.5 ft)

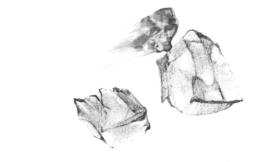

about 95 million years ago: carnivores and primates would have had one, but not the "archaic" mammals predating them.

Today—and let's hope it's not forever—pollution has had a devastating effect not only on the environment in general (soil and food, for instance) but also, more concretely, on the reproductive apparatus. A study examining nearly three hundred polar bears has shown that certain chemicals (polychlorinated biphenyls, which have been banned since 2001 but remain present in the sediment) shortened the size and weakened the density of the bears' penile bones, thus increasing the risk of fracture during mating.

PHALLOSTETHUS CUULONG

—

A Literal Dickhead

To ensure fertilization, some males remain clinging—and sometimes stuck—to the female until ejaculation occurs. One doesn't have to look very far. Dogs have a penis equipped with two protuberances at the base: the *bulbus glandis*. These enlargements of the corpus cavernosum swell in the act of penetration and wedge the penis into the female's genital tract for five to sixty minutes. It's quite handy for keeping others at bay.

But there are better means for hanging on—or maybe worse. Allow me to introduce *Phallostethus cuulong*, a fish about 2.5 centimeters long that was discovered in 2009 in the brackish waters of the Mekong Delta in southern Vietnam. The distinctive signs of this diminutive creature are a translucent body—and genitals right behind the mouth! What a charmer. To be more precise, its *priapium*, or copulatory organ, is jaw shaped. *Phallostethus* means "penis chest" in Greek. There's nothing better for hooking up with the ladies. Needless to say, the goal is to sire as many offspring as possible.

Size of sex organ: about 4 mm (0.15 in)
Size of animal: about 2.5 cm (1 inch)

This priapium is an external organ, and it has many parts—like a Swiss army knife—that have evolved from the pectoral and pelvic fins. The standout features are its shaft and serrated hook. You'd think they would spoil the mood . . . Sure enough, the male uses this bony hook to grab the female's head and force her to stay pressed against him. Love is a many-splendored thing. By holding his partner close, the male has no difficulty getting his sperm into her urogenital opening, which is likewise located on the head. This tête-à-tête lasts much longer than mating in other fish, which is practically instantaneous. Also in contrast to other fish, fertilization is internal, which means that little sperm gets lost, and all eggs have a chance. It's not pretty, but it gets the job done.

We know that much—but still don't have a clue what evolutionary process led this fish to have its reproductive organs in its head.

THE COWPEA
SEED BEETLE

(Callosobruchus maculatus)

—

A Penis Riddled with Thorns

Some animals use penile bones, a *bulbus glandis*, or jawlike organs to hold on to females; others do even worse. Sensitive readers may wish to avert their gaze. Ready? Let's do this. "You're going to have to be strong. Very strong. In a word, I won't mince words. I'll get straight to the point. I'll speak frankly. I'm going to speak frankly. I'm about to speak frankly to you," as the French comic Coluche said when sharing bad news.[2] Well, the male cowpea seed beetle actually pierces the female's body while mating; she endures a real ordeal. The end (reproduction) is supposed to justify the means, but that's what he does. To human eyes, it looks like torture.

These small beetles, native to Africa, have an aedeagus outfitted with spikes. As occurs among members of some other species (for instance, *Bembidion*

---

2 "Le cancer du bras droit," 1976.

carabid beetles and Sepsidae dung flies), the "penis thorns" damage the female's genital tract. Some researchers deem this a way for the male to hold on to his partner (if "partner" is even the right word), but others think it serves to make her genitalia unavailable to other males who come along later. Needless to say, for the war of the sexes to escalate to this point, rivalry among males must be fierce to begin with. But what might be bad for the individual benefits the species, and sure enough, males with the longest spines fertilize the greatest number of eggs.

All of this sounds rather awful, but things take an exciting turn, as some female beetles have evolved to protect themselves from injury. Thus, in populations with a high rate of big spines among the males, females have developed reinforced genital tracts. *Callosobruchus maculatus* also have powerful hind legs, which allow the aggressed party to get rid of her assailant with a violent kick. All the same, once she's been impregnated, her life is shortened: ten days instead of a month.

The situation is perplexing. That said, we should note that spiked penises are not the province of insects alone—not at all. Some mammals have penises outfitted with small, keratinized spikes that seem effective for stimulating the female and ensuring—albeit with some discomfort—that sperm is optimally implanted in her genital tract. This is the case for cats and other felids (lions, for example). In turn, porcupines have a hook at the base of the glans, which stabilizes the penis during penetration—and causes injury to the female upon removal. It seems the purpose is much the same here as among the insects. And insult compounds injury: hurting the female will "discourage" her from mating with others. Love is hell.

Some ladies get their revenge, though. And the dish is served hot. Female *Nicrophorus vespilloides* beetles are extremely aggressive. Gestating and providing for offspring come at a high price for this necrophagous species. Females select partners on the basis of how much food their corpses will provide the young. Once the female has chosen a male, she takes great pains to force him to remain monogamous. The female will even bite him so that he stops producing pheromones attractive to her rivals. But wait, it gets worse. Let's get straight to the point this time: once fertilization has occurred, she secretes a gas that destroys his sex drive. Chemical

Size of sex organ: about 0.9 mm (0.035 in)
Size of animal: about 4 mm (0.15 in)

castration forces him to stick around and help her feed their offspring. And here's another trick: if some guy shows up and tries to mate with her when she doesn't want to, she can change the shape of her copulatory orifice—that is, make it impassable. The evolutionary effects are profound: males struggle to keep up, and over ten generations, the size and shape of the species' genitalia as a whole are transformed. As research on female copulatory organs is still forthcoming, we can be sure the story doesn't end there!

THE RONDO BUSH BABY

(Galago rondoensis)

—

A Prickly V-Shaped Penis

Penises are often extremely complex, and they can vary greatly from one species to another—even within the same family. This is the case in our own family, the primates, especially when it comes to galagos, or bush babies. These distant cousins of ours, whose ranks include fourteen species, are lorisiforms and closely related to lemurs (lemuriforms). Small African primates, bush babies are nocturnal and move around by jumping and swinging in trees; chimpanzees are known to hunt and eat bush babies on occasion. They're particularly interesting to study because their classification has long been the subject of controversy; research gaps in skeletal measurements and the coloring of coats make precise identification uncertain. At any rate, we do know that the penile morphology of nocturnal primates varies considerably. Further work on their reproductive organs will represent a major asset for identifying and classifying species.

Galagos' penises present distinctive characteristics that vary from one kind of the animal to the other. Among other things, these penises vary in shape; they have hard, keratinized spines; and the size and relative position of the baculum (the penile bone, or "love dart," you will recall) vary. The spines may be simple (small and pointed, of medium length), robust (enlarged structures with a single point, often thicker at the base), or complex (displaying several points). One study of 225 bush babies confirmed that morphological criteria of the penis could be used to determine the species; for example, the average surface area of penile spines varies significantly and distinguishably from one species to another.

Galago rondoensis, pictured here, stands out from other bush babies. Native to the mountains and forests of Tanzania and Kenya, this small primate has a penis with a morphology different from all other galagos: large, long, and narrow at the base, it spreads out and assumes the shape of a V at the end. What's more, the baculum—which matches this design, of course—is off-center relative to the glans. Finally, the spines of the penis are clearly bigger than those of other species, but they are mere "buds" when the animal is immature. So the morphology really is quite distinct. *Galago orinus* (the Uluguru bush baby, or mountain dwarf galago) also sports equipment like this. Some researchers wonder whether these two creatures should really be considered galagos at all. Inspecting penises can lead to unexpected results.

But why does morphology differ so much when species are so closely related? For one, there's the matter of sexual selection. Female galagos mate with several males over the long breeding season. This gives females a choice between suitors, on their own terms—say, the sexual stimuli that potential mates provide. On the basis of the freedom to pick and choose, the morphological traits of the males selected are passed on from generation to generation—hence a range of penile characteristics. We can also explain penis morphology in terms of sexual behavior: it would seem that small spines make intromission an easier matter, while large spines are associated with multiple ejaculations. Either way, the ladies have the last word!

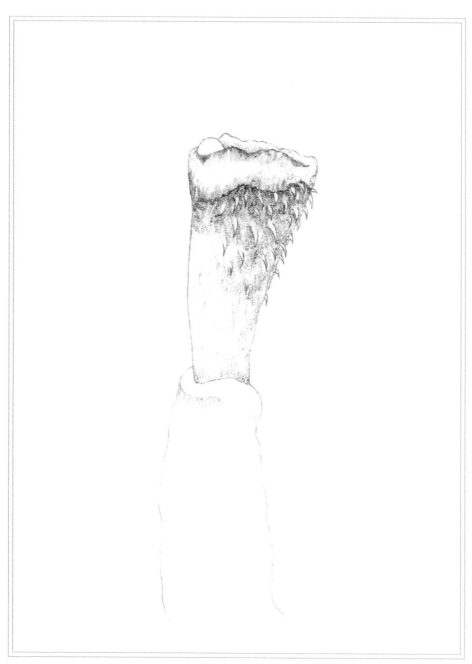

Size of sex organ: about 2.2 cm (0.85 in)
Size of animal: about 40 cm (15.75 in)

THE BEDBUG

(*Cimex lectularius*)

—

Perforate to Procreate

Onward in the world of traumatic extragenital insemination. The torch has been passed to bedbugs and related Strepsiptera (parasitic insects). So what's up here? First of all, male bedbugs (*Cimex lectularius*) are equipped with a beveled penis sharpened into a point that enables them to penetrate the exoskeleton of females. When mating, they pierce from all sides: on the head, belly, legs, back, and even the heart. (Female bedbugs are covered with scars, needless to say.) Then the male injects his semen; only now does the sperm make its way, from outside, into her circulatory genital tract to reach the ovaries.

Why has such seemingly barbarous behavior developed? Probably to circumvent the problem posed by the mating plug left in the genital tract of many female insects. By this means, well-endowed males can also cut across both the female's genital tract and her spermatheca. Allow me to explain. Like leeches and some spiders, bedbugs have multiple "piercings"—as if their vagina extended all over their body. To be more precise, the part of the abdomen where the male deposits his semen, the *spermalege*, is a reproductive system covering the female's entire body, a network that leads to the central sperm bank; in turn, the sperm travels through a system leading to the genital canals and ovaries, bringing it as close as possible to the eggs and maximizing the chances of fertilization.

But that's not all the discomfort that females endure. In fact, "discomfort" puts it too mildly. For some context, males can copulate up to two hundred times a day—a condition known as priapism. In fact, they have trouble telling the difference between other males and females: 50 percent of their encounters are homosexual, 20 percent with completely different animals, and just 30 percent with females. It boggles the mind, but in same-sex contact, the sperm goes down the other party's plumbing and joins the local population. When this passive male takes on an active role and perforates the female, he winds up injecting not just his own spermatozoa but also those of the guy who penetrated him. Some bugs—for instance, *Afrocimex constrictus*, in Africa—have mutated as a result, and males are born with a small, sterile vagina on their back. The goal is to encourage homosexual contact.

Male bedbugs are evidently into sadomasochism. Unless there's some advantage to mixing sperm . . . But we should be aware of the consequences that repeated sexual assault holds for females. To make a long story short, the wounds inflicted can become infected; they reduce the longevity of females and increase mortality by 25 percent. The old saying is that love hurts, but here we see that sex sometimes kills. Meanwhile males have every interest in copulating as much as possible and transmitting their sperm to the greatest extent possible because most of "their guys" are destroyed by the female's immune system. A phenomenal amount of sperm is required for some hundred of these gametes to reach the ovaries; on a human scale, the quantity corresponds to thirty liters for each ejaculation.

So males have more than one arrow in their quiver for optimizing fertilization. But watch out—the ladies don't take it lying down. To escape the abuse and prevent penetration, they have developed an increasingly rigid and elastic abdomen in the course of evolution. Specifically, the spermalege contains resilin, an elastomeric protein that allows the wound inflicted by the male

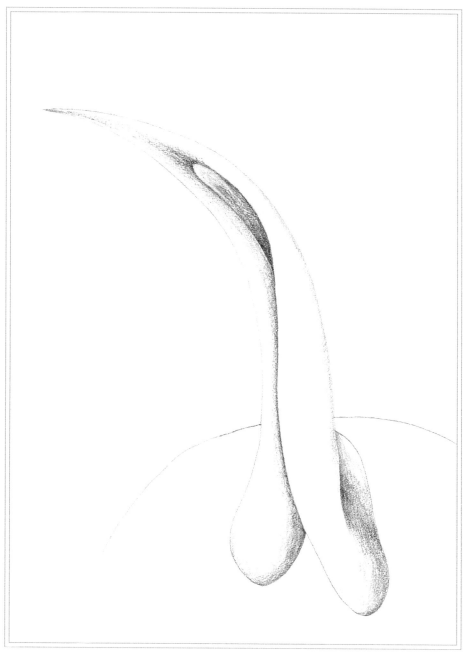

Size of sex organ: about 1.5 mm (0.06 in)
Size of animal: about 6 mm (0.25 in)

to close quickly and reduce the loss of hemolymph (blood, for insects). This means that insemination proves less traumatic. Amazing! Truth be told, it's really more a matter of tolerance than resistance. Insemination is still an ordeal, but females manage to be less traumatized by reducing the energy expended for self-defense. This adaptation benefits both sexes, then, and promotes the survival of the species. But let's not get ahead of ourselves. The males have already retaliated by developing an increasingly pointed penis: another fine example of coevolution. It goes without saying that males with the sharpest penis see the most action.

Finally, there does seem to be the prospect of something like relative calm among bedbugs. Some tropical species are equipped with a penis that's basically a cannon. That's right. They take aim at the vagina, on the female's back, from a few centimeters' distance. The blast is so powerful that the sperm goes in all on its own.

BARKLICE

(Neotrogla sp.)

—

The Females Strike Back

We've already seen how females of *Nicrophorus vespilloides* chemically castrate males to make them monogamous. The barklice of the *Neotrogla* genus use an even more radical strategy. And you thought it couldn't be done.

These barklice are tiny flies about three millimeters long. They live in dry caves in Brazil, where they seem to feed on guano and bat carcasses. Nothing very exceptional. But here it comes. This is a unique case (and who knows, maybe others will be found) of sexual inversion. That's right: the genital organs of males and females work the other way around. We know all about males penetrating females, but here the females pierce the males. You couldn't ask for a more complete sexual revolution.

So what does all of this look like? It's pretty simple: females have a kind of false penis, known as a *gynosome*, which enters the male's body. Is the point to hurt him? Not at all—it's to get nutrients and sperm. If you want something

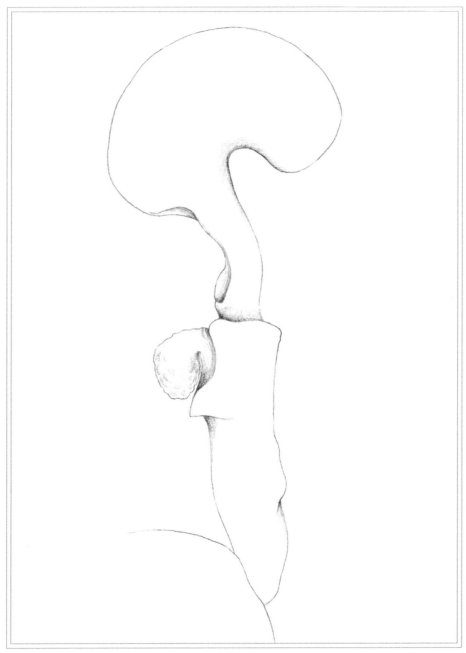

Size of sex organ: about 0.5 mm (0.02 in)
Size of animal: 3.5 mm (0.1375 in)

done right, do it yourself, as they say. And in so doing, Ms. Barklouse takes her sweet time: mating goes on for between forty and seventy hours. That's what it takes for the female to assume a position on top of the male, insert her gynosome, and let the swollen organ hang there anchored by spines in the male's genitals. To get a better idea, know that her false penis sits so firmly that if you try to unhook her, the male's abdomen will rip right off. Nutrients obtained by this means are vital for her to conceive eggs. Such behavior might have developed because the caves where these barklice dwell offer scant resources. If so, this is a way of killing two birds with one stone, so to speak. Just consider the males as sperm banks and refrigerators all in one. That's really turning the tables!

Is anything like this known to happen elsewhere? When it comes to consuming resources that belong to the male, other examples exist, which are just as spectacular—if not more so. For instance, the female sagebrush cricket (*Cyphoderris strepitans*) eats her partner's fleshy wings in the process of mating. But the Palme d'Or probably goes to mosquitoes of the Heleidae (Ceratopogonidae) family. When mating, the female simply pierces her partner's head with the pincers on her mouth. By this means, she can inject digestive juices so that his body will dissolve before she devours it. Incredible. Then she throws away his empty shell, keeping only the genitals (which are now attached to her) for fertilization. It bears repeating: she liquefies her partner, gulps him down, then keeps his stuff for herself. That's really something else.

V

CALLS OF THE WILD: COMMUNICATION, INTIMIDATION, AND DECEPTION

THE AFRICAN SPURRED TORTOISE

(Centrochelys sulcata)

—

An Imposing Organ

So far, we've discussed the functional aspects of the penis (navigating foreign terrain, preventing competition, penetrating in any number of ways) and, on occasion, those of the vagina (self-protection, warding off intrusion, and piercing the male). Let's turn to uses that are less well-known: how penises serve to communicate, intimidate others, or even trick them.

Land turtles will be our guide to this underexplored (and somewhat unexpected) field of inquiry. At first glance, these animals appear to be pretty easygoing. They move slowly and give the impression of being relaxed and calm—almost affable. Well, relative to their size, they have a huge, rigid penis and a glans that swells with blood. So sometimes it gets *really* big—so big, in fact, that they use it to threaten competing males and impress the ladies. Male *Homo sapiens* aren't the only ones who fight over penis size, it seems. The African spurred tortoise (*Centrochelys sulcata*) is a case in point. Inhabiting the arid savannas of the Sahel, these creatures dig burrows ten meters long and three to four meters deep to protect themselves from the intense heat of the day and the coolness of the night. Their burrows are impressive, but their sexual behavior steals the show.

The African spurred tortoise is the largest tortoise on the continent, and second largest on the planet, right after the giant tortoises found on the Galápagos Islands and Seychelles. This species is also known for the confrontations that occur between males. During the mating season, they rumble until the loser winds up getting thrown on his back. Sometimes, however, when fighting for female attention, these gentlemen manage to avoid coming to blows. Other animals might show their teeth, bristle their coat, stand up tall, or throw things to intimidate rivals, but these options aren't available to turtles. Land turtles like our African spurred tortoise have an extraordinary asset for intimidation: a massive penis crowned by a protuberance that's also nothing to sneeze at. The organ lets competitors appreciate the turtle's good health and vigor. It's a way of avoiding a direct physical altercation that falls in line with other forms of communication the tortoise employs: bites, head movements, chemical signals, knocking shells . . . (Courtship displays also include a diverse array of gestures: blinking, nodding, scratching, and so on.)

Nota bene: the penis also serves the purpose of communication among some primates. Baboons and squirrel monkeys, for instance, will sometimes strike poses while sporting an erect member. Evidently such behavior serves to warn of imminent danger or threaten predators. Protect your family and loved ones with your penis!

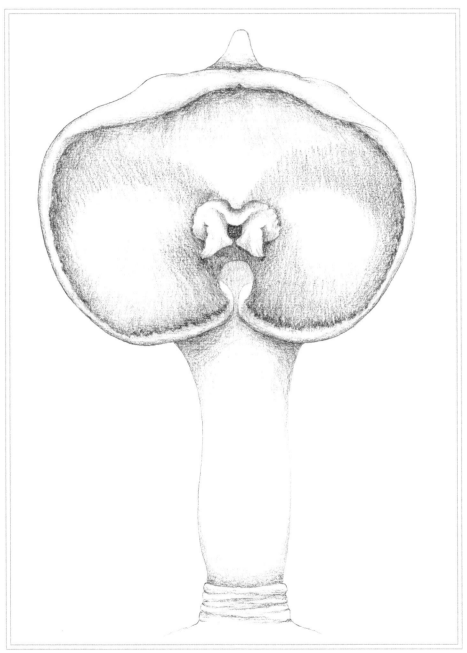

Size of sex organ: about 25 cm (10 in)
Size of animal: about 55 cm (21.5 in)

THE WATER BOATMAN

(Micronecta scholtzi)

—

The Singing Penis

Let's return to the enchanting and wonderful world of bugs, specifically boatmen, water insects that inhabit pools and ponds. This time, communication—and not traumatic insemination—is at issue. The males of this species attract females by singing. Fine, you might say, they're not the only ones who do so. True, but these gentlemen do it with their penis! A penis that makes sound? Yes, indeed. All of two millimeters long, the males rub their organ against the rough ridges of their abdomen to produce a powerful chirping that attracts the females. Sometimes this is called a "singing penis." It's quite charming. But let's not rush to any conclusions about the origins of song in the male genital apparatus!

Attracting females in water poses a major problem: the wavelengths of sound in this element are long. For a potential partner to hear anything, mating calls need to reach a high volume to increase the range of the signal transmission. At the same time, the size of these insects' body imposes a limit, and there's also the danger of attracting predators. This is why male crickets and boatmen avail themselves of resonance structures that produce songs particular to each species, at a volume commensurate with their physical dimensions, which predators can't pick up. Despite their diminutive stature, these tiny insects can really belt it out. Their "singing penis" might be only 50 microns long, but they hold the record for the volume of sound emitted by any insect. It's the loudest penis in the world! The chirping sound can reach 99 decibels. If you don't know what to make of this figure, it's as loud as listening to an orchestra from the front row. For reference, when an elephant sounds his trumpet, it's just over 110 decibels loud—but an elephant is more than 2,500 times as large. Evidently, then, our little bug is the most effective noisemaker, in terms of acoustic energy relative to size, in all the wild kingdom. We've known for some time that the males pitch woo by applying friction to their genitals, but no one suspected that this signal is audible from the water's edge, at a distance of several meters, because it passes through the interface separating the water from the air.

More fascinating still, why such a powerful song has come to be is a mystery, and the precise mechanism at work remains unclear. We must conduct more research in aquatic environments, especially on how the acoustic behavior and anatomy of these creatures have adapted to specific milieus. The intensity of their song qualifies as a secondary sex characteristic, which appears at the onset of sexual maturity—like antlers on deer, the songs of some birds, and changes in color among any number of animals. In this light, we can assume that the boatman's song responds to strong pressure among males to find a mate: during courtship, competition prevails. The most winning sound will lure one or more partners. We need to do further work to find out how the females evaluate the relative attractiveness of singing penises. Does size—volume, that is—matter?

Size of sex organ: about 0.05 mm (0.002 in)
Size of animal: about 2 mm (0.075 in)

THE GRASS MOTH

(*Syntonarcha iriastis*)

—

A Wily Penis

Lepidoptera (butterflies and moths) use a wide variety of means to produce sound. Nocturnal species are known to emit songs inaudible to the human ear. In various ways, they strike their thorax or wings; males of certain species make sounds by rubbing a specialized scale file on their genitals against parts of the abdomen adapted for this purpose. One such example is *Syntonarcha iriastis*, a small moth of Indo-Australian origin belonging to the Crambidae family.

Males perch on trees and bushes, spread their wings, and expose their genitals. Evidently, this exhibitionistic display—their reproductive organs comprise files, scrapers, and resonance zones (among other things)—engages the mechanism that brings forth ultrasonic frequencies. Using a device originally designed for bats, researchers have detected the sounds at a distance of twenty meters. Needless to say, one assumes they're calling on the ladies to mate. But here is where things really get interesting. In fact, it is possible that these songs serve a different function altogether. Some of them strongly resemble the bursts of echolocation that insect-eating bats use to locate prey. Researchers have no doubt that imitating the sounds made by bats is a trick. The exact purpose of doing so is more enigmatic. Does it happen so that the female, sensing a predator's approach, will freeze and stay put—making her available for suitors? It's not out of the question.

Another hypothesis holds that the sounds emitted by this moth serve to scramble the sonar of the bats, making it harder for them to locate their prey. Once upon a time, I wouldn't have been inclined to believe that noise made by genitals could be used to send predators down the wrong path. But nothing that goes on in in the animal world surprises me now—especially since there's no evidence that the females of the species are drawn to these ultrasonic transmissions at all.

Finally, scientists have proposed a third hypothesis, which was developed by studying the yellow peach moth (*Conogethes punctiferalis*), also depicted here. This time, competition between males is the issue. It would seem that they can shoo off rivals with series of short impulses and woo females with sustained notes. Indeed, staccato emissions resemble the sounds bats make, while long ones prompt interested ladies to lift their wings as a sign of welcome. *Dulce et utile*, as classical wisdom would have it.

Size of sex organ: 2 mm (0.075 in)
Size of animal: about 3.5 cm (1.375 in)

THE FOSSA
(*Cryptoprocta ferox*)

—

A Thorny Clitoral Deterrent

The fossa, or *cryptoprocta ferox*, is the largest carnivore in Madagascar. Primatologists know the creature as a major predator of lemurs—including the tiny mouse lemur now being studied here at our MECADEV laboratory. Like lemurs, fossas are threatened by the dramatic decline in biodiversity on the island they inhabit. But that's not why I mention these magnificent animals here; what I'm interested in is their extraordinary sexual adaptation. The fossa is one of the rare mammals among whom females, while still immature, undergo a passing phase of masculinization. What's this, you ask? For a certain period, they make others think they're male! This trick is quite handy for communicating false information and thereby ensuring the females' own safety. So how do they do it, and who—or what—are they protecting themselves from?

For the phase in question, the clitoris grows in size; then the extended organ develops spikes that make it look as if it belongs to an adult male of the species. Incredible! It seems that the girls deck themselves out with this feature to avoid aggression on the part of guys looking to mate; the thorny clitoris acts as a deterrent. In brief, transient masculinization enables young female fossas to avoid sexual harassment (don't forget what female barklice do) and improve their prospects for survival. Unwelcome attention from males, after all, can lead to injury or even death. The fact that mature females aren't always around and the annual estrous cycle doesn't last for long makes males try to mate quickly and frequently—hence their aggressive behavior. Juvenile females are especially vulnerable because they're small and have just left home, as it were. The disguise effected by clitoral transformation enables them to escape detection by males. Plus, genitals like this pose a barrier to coition. Quite the radical solution!

At the same time, masculinization serves to protect juvenile fossas from others of the same sex. Danger is everywhere. While males seem somewhat territorial—data from radio tracking are limited—mature females have areas they consider entirely their own. At the age when they are just starting to leave the home, young lady fossas haven't yet found their place in the world and risk unfortunate encounters with older females. An enlarged clitoris helps on two fronts.

Let's be honest: compared to what we know about penises, we know nothing about vaginas and clitorises in mammals. Each new discovery promises a revelation about connections between morphology and sexual behavior. For instance, female spotted hyenas (*Crocuta crocuta*) have a clitoris that does just about everything. Sometimes called a "pseudopenis" because of its large dimensions—which are readily apparent—it is used for mating, urinating, and . . . giving birth. Evidently, this long, sinuous, winding genital apparatus enables the female to select the best spermatozoa from all the partners she encounters. Clearly we need to make more funding available for work on vaginas and clitorises.

We have reviewed numerous functions of the penis, vagina, and clitoris. If the overall balance seems uneven, it reflects the fact that the vagina and clitoris remain uncharted territory. At any rate, we've seen penises used

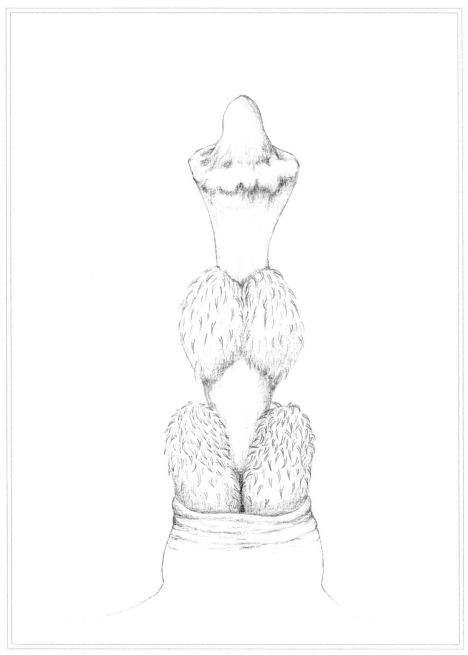

Size of sex organ: 10 cm (4 in)
Size of animal: about 70 cm (27.5 in)

for penetration, navigation, plugging up the female's genitals, cleaning them out, holding on tight, mutilation, piercing, stimulation, and even communication. We've also observed how the vagina and, to a lesser extent (for want of information), the clitoris can penetrate, reinforce violated areas, block off intrusive semen, store sperm, digest it, and communicate. Needless to say, all these impressive capacities serve the purpose of survival. But where does pleasure fit into the picture?

VI

PLEASURE SEEKERS

THE CAPE
GROUND SQUIRREL
(*Xerus inauris*)

—

Handy Masturbation

We have reviewed a multitude of behaviors and adaptations, each one more impressive than the last and always advancing evolutionary interests: to reproduce and perpetuate the species. But we shouldn't forget that many animals—not just human beings—experience pleasure, and that sex sometimes has functions other than fertilization. Just for fun. Although little has been written on orgasms and pleasure, they also exist in the animal world. Often it's a matter of consent, with preliminary arrangements to prepare both partners for the act itself.

If one behavior in particular shows that pleasure alone is the goal, it's masturbation. Ameline Bardo, my dear former doctoral student, has observed great apes, chimpanzees, bonobos, and orangutans pleasuring themselves—and using "sex toys" (sticks, for example) to do so. One might argue that these animals are closely related to human beings, of course. However, masturbation occurs just as much among elephants, koalas, horses, and even kangaroos and porcupines; these creatures will stimulate themselves by friction to the point of ejaculation. The benefits of this practice would seem to be many.

A diminutive rodent provides the perfect illustration. Masturbation often has a pejorative connotation. I'm pleased to be able to show that a creature as adorable as the Cape ground squirrel (*Xerus inauris*) can indulge in what used to be called "self-abuse" without losing any of its charm. Without further ado, allow me to present this cute little inhabitant of the arid regions of southern Africa, better known for employing its tail as a parasol to regulate internal temperature under the blazing sun. One can readily find photographs showing Cape ground squirrels standing on their hind legs on the lookout for predators, like meerkats (with whom they sometimes even share a burrow).

Males and females of the species live in separate groups. That's probably not enough to explain the practice of masturbation, but it is worth noting. Cape ground squirrels reproduce all year round, with a peak in winter. Another point warrants mention: males and females alike have multiple partners; on this score, too, there's no obvious reason why they would seek any short-cuts to reaching orgasm . . . These are just the facts. Here's another one: the male's testicles take up about 20 percent of the body's whole length. Impressive, I daresay.

These points of interest, taken together, just might indicate why male Cape ground squirrels masturbate. But while it's clear (as we will see a little later on) that rodents can experience orgasms, and therefore pleasure, it seems that masturbation among these squirrels also has a highly specific function: to help them avoid getting sexually transmitted diseases. If so, masturbation would serve as a kind of sexual hygiene. In a three-hour estrus cycle, a female Cape ground squirrel will mate with up to ten males. By masturbating after copulation, males can reduce the risk of infection. That's not nothing, in light of the fact that sexually transmitted infections can affect fertility significantly. STDs likely influence these little creatures' mating strategies. So here we have a case of self-medication, as it were—or, more precisely, a healthy dose of self-protection. Self-love is safe sex.

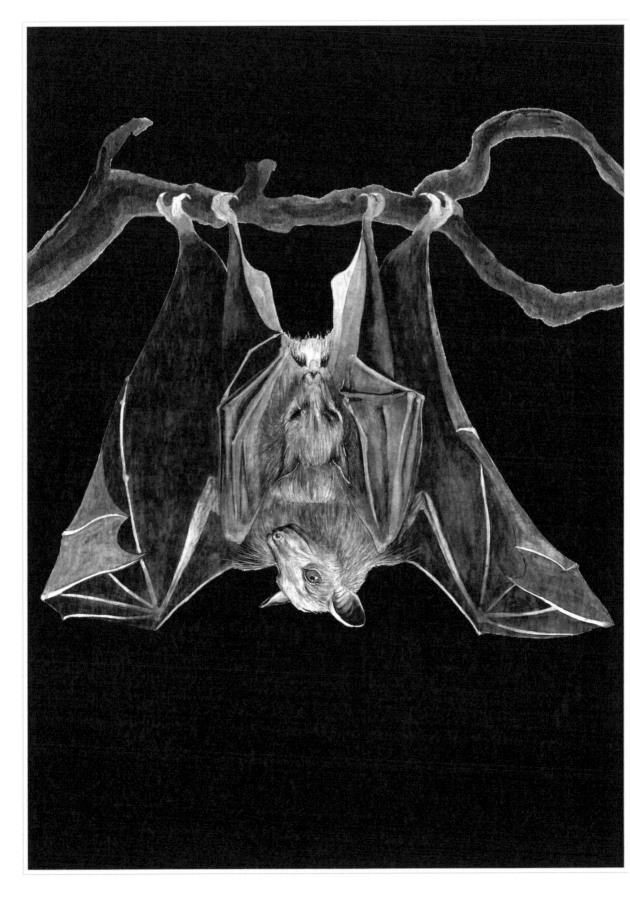

THE GREATER
SHORT-NOSED
FRUIT BAT

(Cynopterus sphinx)

—

His and Hers

Bats (chiropterans) are flying nocturnal mammals; rarely touching the earth, they use echolocation, ultrasound, or sound and smell to orient themselves, detect prey, and avoid obstacles. They're the only mammals known to fly actively—in contrast to flying squirrels, members of Phalangeridae, or colugos (*Galeopithecus*), who all glide on "wings" formed by a membrane connecting the body, limbs, and claws. Bats at rest make quite a sight, hanging upside down from their toes. Of course, what interests us here is their mode of reproduction. For many of these species, fertilization occurs at staggered intervals. Females store sperm in their genital tract; ovulation and fertilization happen later. So far, no pleasure.

Fruit-eating bats—especially rousette bats—avidly perform fellatio and cunnilingus. It might be hard to picture these little critters engaging in such behavior, but it's real enough and has received serious attention from researchers. Thus, studying one colony, scientists observed that most instances of penetration were preceded and followed by cunnilingus.

Regarding cunnilingus, by the way, I can't help but share a discovery concerning spiders (and tip my hat, again, to my colleague Christine Rollard, the arachnophile). One in particular, Darwin's bark spider (*Caerostris darwini*), is known for spinning huge and extremely tough webs—and for displaying less violent sexual behavior than many other species. I trust the reader won't have forgotten our examples of females emasculating their partners and other genital mutilations. Well, none of that goes on here. True, the male might attempt to sneak up and copulate by surprise or try to hold on to the female by tying her up in his web . . . but what's really remarkable is that he can perform oral sex before, during, and after mating! Bet you didn't see that one coming. Apparently, by depositing saliva on the female's genitals, the male bark spider creates a chemical environment favorable to fertilization, thereby enhancing the likelihood of paternity.

But let's get back to bats, and specifically to the greater short-nosed fruit bat (*Cynopterus sphinx*) illustrated here. Most females of this species, found in Asia, fellate their partners while mating. Flying around or back at home, these animals know how to have a good time. To get technical, the female will tilt her head and lick the base or shaft of the male's penis (not the glans, as it's otherwise engaged). It's a feat worthy of the *Kama Sutra*. Fellatio continues for the duration of intercourse; as long as she keeps going, so does he. In fact, one second of oral attention translates into ten more seconds of mating; sex including fellatio always lasts longer for these animals.

Needless to say, "pleasure" doesn't qualify as a full explanation. Researchers have suggested that licking the genitals provides lubrication and helps prolong the time of penetration. Since saliva has antimycotic and antibacterial properties, it's also possible that fellatio can prevent sexually transmitted infections. Finally, the activity may play a role in choosing a partner to begin with, relaying chemical signals (that is, molecules present on the penis) to a system of recognition known as the "major histocompatibility complex." Complex, indeed, but fascinating—as always. Incidentally, fellatio among several great apes seems to serve the main purpose of giving pleasure.

THE BOTTLENOSE DOLPHIN

(Tursiops truncatus)

—

Homosexual Role Play

Although sexuality evolved primarily for reproduction, it takes many forms. Males generally have sexual relations with females, but not always; numerous possibilities exist. Nonreproductive sexual behavior has been observed between individuals of the same sex in many species. Like many other forms of behavior, homosexuality isn't unique to humans: more than a thousand species exhibit homosexual—or, more precisely, bisexual—traits, and these species include mammals, birds, reptiles, amphibians, fish, and even insects.

Among birds and mammals, homosexual behavior that involves sex, courtship, and co-parenting may be observed in geese, flamingos, gulls, oystercatchers, warblers, deer, zebras, giraffes, gazelles, sheep, elephants, manatees, skunks, rats, chimpanzees, dogs, bulls, Humboldt penguins, ducks, and others still.

At least ninety-three species of birds are in the club. Fifteen percent of male gray geese and 20 percent of seagulls are strictly homosexual, and black swans sometimes form homoparental pairs. Same-sex relations are especially pronounced in primates—including humans, of course. For more distant species—squirrel monkeys, for example—same-sex relations appear to be restricted to games and contests for dominance. Among apes closer to man (macaques, baboons, chimpanzees, gorillas, bonobos), homosexual behavior is a more complex phenomenon that involves friendship, gestures of reconciliation, regulating tensions, and forming alliances.

At any rate, homosexuality and bisexuality are common in the natural world. Scientists now accept that sexual relations can often serve more than reproductive purposes, but it took some time to get here. In 1884 at the Sorbonne, Abbé Maze, a French clergyman, communicated his observations concerning "pederasty" among June bugs—or cockchafers, as they're also known. A few years later, a distinction emerged between two kinds of "pederasty": when it occurs because of a lack of females, and cases in which it's a matter of preference, or taste. So some progress is possible, even though prejudices can prove tenacious.

The bottlenose dolphin (*Tursiops truncatus*) is no exception. In fact, from a prudish human perspective, dolphins are real libertines, known to masturbate with dead fish, "mate" with buoys, and perpetrate sexual harassment . . . You'll never watch *Flipper* the same way again.

Tursiops truncatus is a toothed cetacean (odontocete). The object of intensive study both in captivity and in its natural environment (along the coast of Florida, in particular), the bottlenose is the best-known species of its family. And that's lucky for us, because these animals are quite the discovery. The most stable relationships in this species are formed by young members of the same sex. Once the males grow older, they will work together to capture a fertile female and mate with her, using some coercion. Such behavior raises questions—maybe boy dolphins train to optimize sexual contact with females when the time is right. That's one hypothesis, anyway. But if that's the case, how do we explain them rubbing their penises against each other? What about anal sex and—get this—"nasal sex," when the penis is inserted into the blowhole?

The upshot is that research on animal sexuality (including the activities of human beings) has documented a vast array of practices, and there's no shortage of things yet to be elucidated—especially the evolutionary role of so much "funny business."

THE BONOBO
(*Pan paniscus*)

—

Anything Goes

Homosexuality—or, more precisely, bisexuality—runs throughout the animal world, then. Females are no exception, especially in our own family, the primates. Among mountain gorillas, homosexual behavior between females is widespread. Their cousins among stump-tailed macaques (*Macaca arctoides*), sometimes called bear macaques because of their coat, engage in an array of homosexual behaviors. For instance, when macaques are in heat—a condition made evident by the red color of their faces—females will rub their vulvas together and emit the sounds typical of mating. But they don't stop there; these ladies are also known to masturbate on deer . . . You read that right. They rub their vulva against the deer's back—which is something macaques also do with other females of their kind. They're quite serious about it, too. One study charted five females performing some 260 mounts of this nature. Indeed, stump-tailed macaques will even compete for a "ride," dominant ones pushing their social inferiors aside (as occurs in mating of the heterosexual sort). Interestingly, whereas the deer go along with the appetites of female macaques, they refuse to do the same for males.

Bisexuality is also the rule among bonobos (*Pan paniscus*). For this species of great ape, sex serves the additional purpose of appeasement when conflict erupts. In fact, it works like magic: disagreements rarely escalate. As the leading primatologist Frans de Waal has observed, chimpanzees resolve sexual issues through power, whereas bonobos resolve issues of power through sex. All on their own, these animals show that sex in the natural world can be completely unrelated to reproduction. Evidently, three-quarters of their erotic life has no direct connection to perpetuating the species. Sex occurs repeatedly, independent of seasons and ovulation cycles. Every day, with a wide range of partners, and in an array of different positions, these great apes make sure to have a good time.

Why do without, especially if it relieves tension? Aren't "peace and love" what it's all about? And since heterosexual mating takes place face-to-face, why not call it love? But maybe heterosexual encounters are too boring. When that's the case, bonobos go for something else—which means pretty much anything at all: French kissing, masturbation, fellatio, "petting." Just for the sake of pleasure, over and over. The facial expressions and sounds these animals make leave no room for doubt—about female orgasm, too.

So there's love for you. Peace is a little more complicated. We should note an element of calculation in the picture. For instance, when a female mates with a large number of males, she is typically protecting her young. How does that work? Well, it means that any one of her partners might be the father. In fact, infanticide does not occur among bonobos—in contrast to other primates. So, yes, love *and* peace.

In sum, bonobos enjoy themselves just like the children of nature they are. Sex represents one activity among others for smoothing out and spicing up social life—a break from delousing or eating, for example. What's the problem with that? I can't see one.

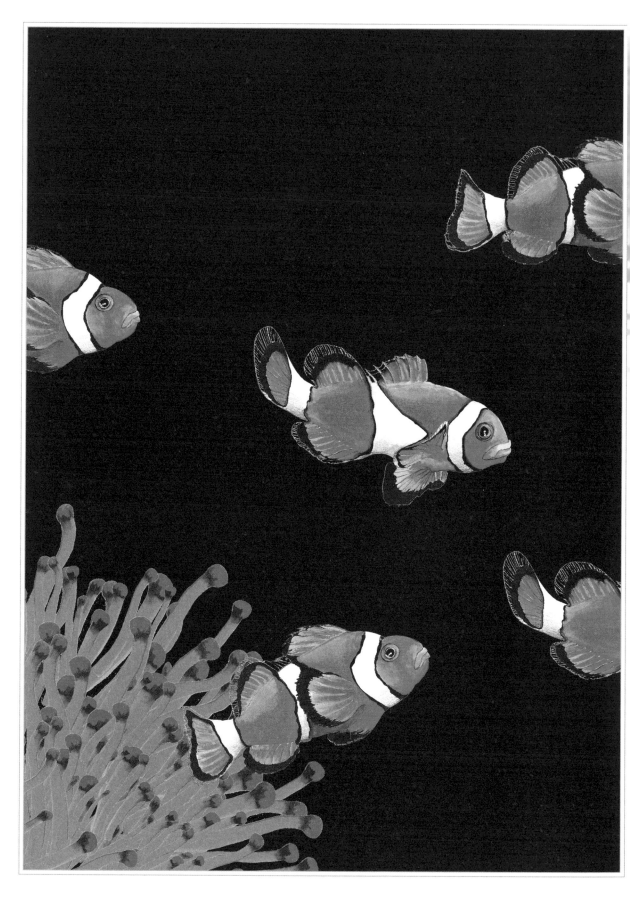

THE CLOWNFISH
(*Amphiprion ocellaris*)

—

Changing Sex and Size

Many animals possess both sexes' genitals at the same time. Earthworms start out as males, then lay eggs and fertilize them after copulating with another male. When they mate, they're the same sex. This is full-on hermaphroditism. But other cases are more impressive still. One wouldn't even suspect as much with some species. You're probably familiar with the film *Finding Nemo* (2003): the adorable hero embarks on a series of adventures when his mother dies. Well, you'll never look at this charming little clownfish (*Amphiprion ocellaris*) the same way again. His real-life counterparts don't hesitate to embrace radical transformation: they change size and sex.

These small orange fish with black and white stripes begin their lives when they hatch from eggs laid on a reef. The newly born larva drifts off into the ocean deep for a few days before returning to its place of birth. From here on out, the transparent hatchling turns into a juvenile boasting shimmering colors and white stripes (with patterns that vary between species). In turn, the young clownfish goes looking for a sea anemone, with which it will spend its whole life living in symbiosis. To be precise, the relationship is a matter of so-called mutualism: the fish passes the rest of its days among the anemone's tentacles and, because it is unaffected by their sting, gains protection; the anemone, on the other hand, gets to feed on the fish's nutrient-rich droppings.

The relationship offers other advantages, too. During the night, when no photosynthesis occurs to provide the anemone with oxygen, the clownfish will move its fins to make sure that water with "fresh air" reaches its host. That's pretty cute. Anemones certainly grow and thrive thanks to these little fish who depend on them. Incidentally, global warming has led some anemones to turn white—and be passed over by clownfish. Neither one of these fantastic organisms will survive if this process continues (and that's not factoring in pollution).

Fantastic organisms? Yes, indeed. And I haven't even mentioned the most impressive feature of clownfish. Their social structure is fascinating. Groups are dominated by a large female who defends the host anemone from tiny attackers. Next comes a male, not quite as large as she is, followed by juveniles arranged hierarchically by size, big to small. If one of the juveniles dies, the next one in line takes its place and quickly assumes the dimensions of the departed; getting any bigger would mean expulsion by the one in front—death, in other words. Whenever a new individual arrives, he or she goes to the back of the queue; the others defend their positions vigorously. Some juveniles wait up to thirty years for their turn as leader of the pack; life expectancy is an astounding fifty years.

More remarkably still, if it should happen that the alpha female gets eaten by a predator, the male who was second in command changes sex and takes charge of defending the anemone! The term is "sequential hermaphroditism," not transgenderism. The first juvenile in the line will become a male, and everybody moves up a spot. Each one increases a little in size, but not too much. Since the increments are reasonable and don't undermine anybody else's position, conflicts are avoided. Everybody will get their turn.

Back to the movies. In *Finding Nemo*, our hero's father takes over caring for his son when the boy's mother dies. In reality, though, daddy would turn into another mommy—and make a new batch of babies with Nemo, the juvenile male. That's an awkward story to tell children, though, and unlikely to receive a G rating.

THE RAT

(Rattus norvegicus)

—

Ultrasonic Orgasms

The following is dedicated to Savonnette, my son's adorable pet rat. It's both poetic and symbolic to end this book with creatures of her kind—charming, intelligent, and noble animals who, unfortunately, are so unloved by humans. These qualities, you may object, don't fit with the subject of the volume at hand. Well, you're right. But rats demonstrate some amazing abilities. And among the surprises they offer, one does relate to our sexual theme: orgasm.

Time and again, sexual pleasure is equated with orgasm. But what is an orgasm? The answer seems basic enough: you know it when you have one. That's true, but how does one define it, so we can see if it occurs in other species? A widespread—and mistaken—view holds that orgasm coincides with ejaculation. This is often the case for men (not always, though), but not for women—even though "female ejaculation" does occur in some instances. Most definitions describe physiological and emotional sensations converging in a moment of ecstasy followed by relaxation: a release of sexual tension that has built up over a period of stimulation and is identifiable by the contraction of the pelvic muscles. If so, orgasm exists in various primate species; the intense uterine contractions and increased heart rate of female great apes during intercourse admit scientific quantification.

From our anthropocentric perspective—and there's nothing wrong with that—we could say that this is only normal. After all, great apes are practically family. The thing is, orgasms happen elsewhere in the animal kingdom, too. Poor, ill-famed rats have them. It's a proven fact! If one focuses on pleasure, research provides an unequivocal answer. Males and females alike show all the signs of truly enjoying themselves; after mating, the male sometimes falls asleep—and the female hops around to wake him up and let him know that she's ready for more. Numerous experiments have shown as much: scientists perfumed males with an almond scent; after a good time with one of these gentlemen, females went looking for others with the same smell!

To resolve the issue with absolute certainty, researchers first inventoried all the physiological signs indicating the approach and occurrence of orgasm in human beings. (That said, it's quite possible that rats show other signs specific to them alone, and we can't detect them.) The cues are many; they include blood flowing to the penis or clitoris, contractions of the perineum, the release of endorphins and opioids in the brain, ejaculation in men and uterine contraction in women, all kinds of vocalizations, then relaxation and detumescence (a fancy word that means the blood is flowing away from erectile tissue). Studies have found all these signs in rats, both male and female: tightening muscles, the release of "reward neurotransmitters" (following the same principle as our endorphins), the emission of sounds—ultrasonic ones, at that—and so on. In a word, there's no doubt about their intense pleasure. This is likely the case for many other species—say, male and female dolphins on their own or together, but also bears, pigs, cats, dogs, and the rest of the menagerie. Everything still remains to be discovered about the pleasure of different species, its origins and evolution. Understanding this aspect of the animal world might even explain why some people have trouble reaching orgasm and help us devise therapies. Here's to love, little rats!

The Rat

CONCLUSION

Gentle readers, curious lovers of nature and friends of animals, we've arrived at the end of our voyage of discovery to the land of genitals. Although the tour was far from exhaustive, we've seen an incredible variety of specimens. Why is there so much morphological diversity? How have reproductive organs evolved over the ages?

And why, in particular, did penises evolve? You'll recall that more than 450 million years ago, fish released their semen and eggs into the salt water, and fertilization could proceed unimpaired. On land, however, the sun would dry out this fragile biological matter. Thus, for the purpose of survival, a new solution emerged: internal fertilization. Animals have spread far and wide over the earth by this means—which involves the penis, the vagina, and, let's not forget, the clitoris. As always, nature has demonstrated great creativity and invented a countless number of forms. Some insects that look quite similar are endowed with very different penises. The same is true of mammals; the males of some species have penis bones, for example, whereas human beings do not.

Why does such fascinating diversity exist? Morphological variety is everywhere. From insects to mammals, from birds to reptiles, penises take on an unbelievable array of shapes. One encounters pincers, tubes, spines, hair, hooks, and more still. All is fair when it comes to spreading one's genes and surviving: piercing and puncturing, self-abuse, plugging up (or scraping out) one's partner, detaching the penis itself, changing sex, bullying, and rape. Reproduction explains, if not justifies, it all. Human ingenuity (or perversity) has invented nothing—not even pleasure, which is shared by more species than we might think. The range of possibilities is certainly impressive. The most widely credited theory is that distinct reproductive organs developed to discourage mating between different species; that said, such a precaution seems unnecessary, since encounters between different kinds of animal, when they take place at all, produce only sterile offspring.

At any rate, diversity clearly thrives at all levels: social interaction between males and females (competition, monogamy, multiple partners), modes of seduction (building bowers, offering gifts, a new coat or plumage, tricks), parade techniques (dancing, singing), mating itself (on the ground, in flight, in various positions), and setting and circumstances (out in the open, in the forest, with predators nearby or absent). There's no reason, then, why morphology shouldn't display comparable variation. Nothing has been set in stone.

So how did genitals adapt and evolve to take on the varied forms they now have? We've managed to get some idea thanks to the examples shared here. Penile morphology is linked to the function and mode of reproduction—mating, that is—of the animal in question. For instance, species in which the female is likely to mate with more than one male display more elaborate genital morphology, which allows the mother to choose the father of her young. The matter is highly complex, of course, and demands further study. In particular, for real elucidation, we need a fuller look at female organs. Research on how reproduction and genitalia have evolved has left the vagina and clitoris by the wayside. It's impossible to understand developments on a large scale without knowing how instances of coevolution

have unfolded—the relationship between changes in both males and females, that is. We still have plenty to do.

Scientific research performed over the course of the twentieth century focused on the male genital organs. Unfortunately, the prejudice only worsened over time—and it's evident in more recent efforts, which have stressed the dominant sexual role of males. Still, new studies have highlighted the speed at which female genital characteristics can evolve, as well as the complex dynamics at work in the coevolution of genital structures. Our understanding of how sexual and reproductive organs came to be what they are is hampered by an outdated but persistent bias. Will female genitalia now receive the attention they deserve? They must! Especially since the few studies that have been conducted are so fascinating. We've only skimmed the surface by noting that females can penetrate males (*Neotrogla*), determine paternity by blocking access to sperm (water striders or boatmen, ducks, some spiders, dolphins), stockpile sperm (some turtles, snakes, ants), digest unwanted sperm to avoid fertilization (some members of Lepidoptera), and repair tissue traumatized by their partners (cowpea seed beetles, bedbugs).

On this note, let's not forget what might be the most important part of all: pleasure. Pleasure comes from many sources in the animal world: homosexuality both male (manatees, skunks, rats, penguins) and female (bonobos, turtles, lizards), fellatio and cunnilingus (bats), masturbation (squirrels, kangaroos, primates, dolphins), sex toys (chimpanzees, orangutans). Many species enjoy sex and experience orgasm without reproducing.

You'll find it all: heterosexual, lesbian, gay, bi, and trans behavior, as well as monogamy, polygamy, and polyandry. Systems of alliance and family organization are intimately tied to modes of mating and therefore to the morphology of genital organs. Under polygamous conditions (which represent the most widespread system), male mammals do not take care of young, and they fertilize as many females as they can. That competition is so rare has likely influenced the shape and size of genitalia. It's certainly no coincidence that male gorillas—living in harems with little rivalry—have very small penises, despite their imposing stature.

In contrast, polygamous species that don't share partners—like chimpanzees—sport hefty organs. They live in groups with several males and females, and the competition is tough. Among monogamous animals—some 10 percent of mammals—parenting responsibilities belong to both partners. The territory to defend is large, and food is scattered. It makes sense, then, for males to team up with one female instead of wasting energy trying to control several of them. Examples include small carnivores (fennec foxes, jackals), some primates (gibbons, marmosets, titi monkeys), and rodents. Tensions can run high, also between females; the diversity of genital organs reflects as much.

Finally, we have polyandry. Females call the shots among ring-tailed lemurs, whose social organization is matriarchal, and community-oriented naked mole rats. Then there's the antechinus—a little marsupial sex addict—whose trysts never end well. During the mating season, the female "gets busy" for twelve hours with as many partners as possible; occasionally, the latter wind up losing their fur and dying from fatigue or internal bleeding. In turn, when the female Tasmanian devil is sizing up suitors, she bites and claws them to find out who is the most vigorous; she gives the weakest a hiding (literally). Once mating has taken place, the male falls asleep—and she trots off to find somebody else.

It follows, then, that genitalia and their uses differ depending on whether the species is polygamous, monogamous, or polyandrous. Examining female organs would help us understand how coevolutionary developments relate to other behavior—overall social interaction, for example. The clitoris has garnered little interest because it's wrongly supposed not to play a role in fertilization. How has it evolved in terms of shape and size? How did it come to have a bone, and why did it disappear in some species? How does it really relate to fertilization, and what is its evolutionary connection to vaginas and penises? These questions pose a major challenge to

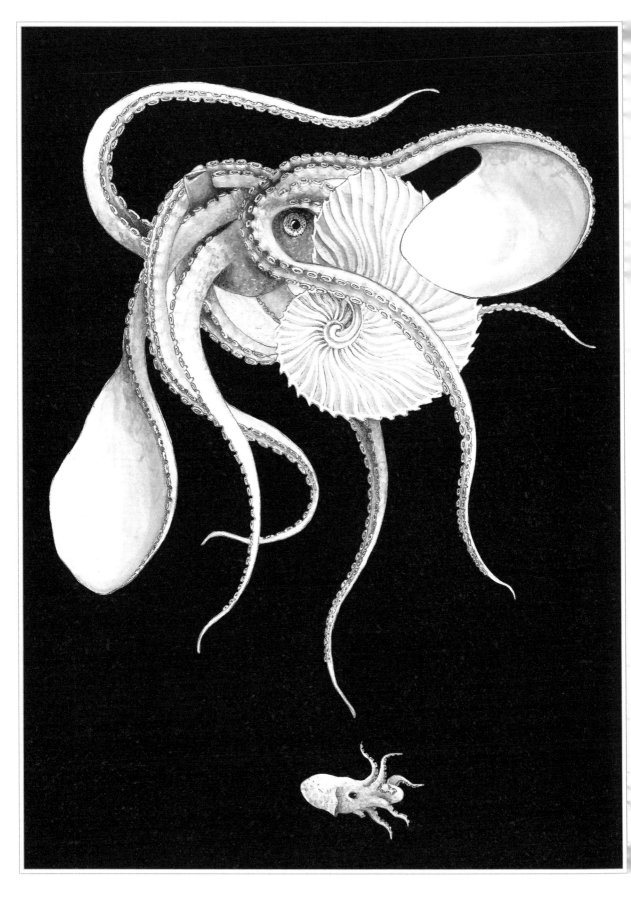

efforts to understand how species survive and evolve on the whole. The book has yet to be written on this vital but forgotten player in evolution and sexual selection, but one of my students has already turned an eye to the matter. To be continued . . .

In conclusion, human beings haven't come up with anything new. Humility is in order here—we're just a drop in the ocean. "Sexodiversity" existed long before we ever came along. As always in the animal world—no matter which part or aspect—a world of discovery awaits. Four billion years of life on this planet have witnessed extraordinary developments. So let's do what we can to make sure it continues. I'm terrified to think these animals might not be around one day, soon or in a more distant future. What will I tell Savonnette . . . or my son? No trees, no bees, no butterflies, no dragonflies, no birds, no squirrels . . . What will there be to inspire wonder? My son and his friends are awestruck by a ladybug—as I once was, too, playing in my grandparents' garden. Or when contemplating those little spiders in my parents' yard. Just looking at an orangutan or elephant—intelligent creatures who return one's gaze and understand—inspires tears of joy and shivering thrills of happiness. Don't let it ever end.

ACKNOWLEDGMENTS

My warmest thanks to Valérie Dumeige for her trust and forthrightness in our exchanges, as well as her spontaneity and readiness to share ideas. Here's to future projects! My gratitude and admiration also go out to Julie Terrazzoni, whose every picture leaves me speechless and full of inspiration. Affectionate thanks to Vivien Boyer, whose presence is always appreciated. Thank you to everyone at Arthaud, especially Isaure Angleys and Karine Do Vale, for their able assistance, good humor, and honesty.

Equally, I would like to thank Ameline Bardo and Marion Segall, former students and now colleagues, who started our research team's game of identifying animal penises a few years ago. Please accept my gratitude for inspiring this book and my most amicable sentiments. I hope to address these weighty scientific matters with them in years to come. Finally, my deepest admiration extends to all the researchers who never cease to amaze me, particularly my colleagues at the MECADEV laboratory.